William Thomas Lyle

Parks
and
Park Engineering

Elibron Classics
www.elibron.com

Elibron Classics series.

ISBN 1-4021-5182-9 (paperback)
ISBN 1-4212-7776-X (hardcover)

This Elibron Classics Replica Edition is an unabridged facsimile of the edition published in 1916 by John Wiley & Sons, Inc.; Chapman & Hall, Ltd., New York ; London.

Courtesy George S. Webster, Chief Engineer, Bureau of Surveys, Philadelphia.

The Walnut Lane Bridge over the Wissahickon. *Frontispiece.*

PARKS

AND

PARK ENGINEERING

By

WILLIAM T. LYLE

Professor of Municipal Engineering at Lafayette College
Associate Member, American Society of Civil Engineers
Member, Society for the Promotion of Engineering Education

FIRST EDITION

FIRST THOUSAND

NEW YORK
JOHN WILEY & SONS, Inc.
London: CHAPMAN & HALL, Limited
1916

THE SCIENTIFIC PRESS
ROBERT DRUMMOND AND COMPANY
BROOKLYN, N. Y.

PREFACE

IN an age of many books every author must make his apology for a new volume. The output of engineering treatises and text-books is so great that for a time at least, in certain branches, further additions should be made only for very good and exceptional reasons. The author of this little book believes that by reason of the dearth of information on the specialized subject treated, a sufficient reason exists for the publication of his manuscript, which for several years in abbreviated form he has used in his classes at Lafayette College.

The magnitude of the park movement in America is only beginning to be appreciated by the average well-informed person. Our larger cities already have their parks, most of them but recently acquired, and it will not be long before all of our American cities will have them also —not merely a few small city squares, but extensive modern parks, reservations and playgrounds, connected by parkways and boulevards.

This book is prepared principally for the benefit of the young and inexperienced engineer of construction. The author hopes that it may be found useful to members of newly formed park associations and commissions; and that its earlier articles may be of help to public-spirited men who may be considering the ways and means of securing for their own communities the great benefits which flow from an ample and well-coordinated park system. It also may contain valuable suggestions to engineers and others who, though not trained in the art, may be engaged in the work of development of private estates.

While the master mind in the conception and design of our largest and best park systems is the landscape architect, it must be understood that the execution of his general plans is peculiarly the function of the engineer, as is also the design of the engineering features. The art of the park engineer, though a specialty, is, however, not a narrow specialty. He must be proficient in matters pertaining to the acquisition of lands, and be well versed in a great variety of engineering operations, such as earth excavation, masonry, water-works and sewerage construction, road building and lighting, and occasionally the construction of steel and reinforced concrete bridges. He must also be a good expert witness.

Wherever possible the author acknowledges the kindness of those who have aided him with advice and information and of those who have furnished him with illustrations for his work.

He is under special obligation to Mr. John C. Olmsted, Landscape Architect, of Brookline, Mass., and to Mr. Howard J. Cole, of New York, formerly Engineer in Chief of the Essex County Park Commission.

WILLIAM T. LYLE.

March, 1916.

CONTENTS

CHAPTER I

DESIRABILITY AND ACQUISITION OF PARKS

CHAPTER II

LANDS AND SURVEYS

CHAPTER III

DESIGN

Chapter IV

LABOR AND CONTRACTS

Chapter V

CONSTRUCTION

ILLUSTRATIONS

PARKS AND PARK ENGINEERING

CHAPTER I

THE DESIRABILITY AND ACQUISITION OF PARKS

THE **Social Need of Parks.** In view of the rapid growth and the increasing densities of population of our American cities, the need of parks, breathing spaces, playgrounds and parkways is becoming more and more imperative. Compact urban development has created a need which should be met by the municipal governments responsible for the condition. Our cities are recognizing the necessity and moral obligation of providing places for rest, recreation and wholesome open-air amusements. Excepting in a few of our larger cities, such as New York, Brooklyn, Philadelphia, Boston and Chicago, the movement is less than twenty-five years old, and even in these places great advances along new and better lines have been made in the last two decades. With new conditions new needs are born, and though our parks have supplied fairly well an urgent need, many of them in the near future will have to be remodeled in order that they may be increasingly useful, not only for an increasing population, but also to each individual in the community in need of the health advantages which it is possible for them to offer.

The Economic Need. The policy of delay in the purchase of park lands, if it may be called a policy, is certain to be an expensive one. As an example of this may be cited the fact that for three parks covering less than ten acres in the congested portion of the East Side, New York recently paid more than it paid for Central Park, which has an area of 840 acres. As another example, Hudson County, New Jersey, has paid for its parks an average of over $3000 per acre, and for some property acquired by condemnation proceedings as high as $23,000 per acre; whereas Essex County, adjacent, has purchased 3000 acres of mountain land at a cost of only $500 per acre. Compelling the present generation to bear the entire burden of purchase is not advocated, but by incurring a bonded indebtedness, the coming generation will bear its share of an expense which, if not incurred now, will increase to such an extent as to render, in many cases, the purchase almost impossible.

The judicious selection and purchase of park lands and reservations is a far-sighted policy, which like the laying out of wide streets, is sure to be financially profitable in the long run.

It is a well-known fact to the expert that some of the most undesirable lands from the standpoint of the builder and real estate broker are the very best from the standpoint of the landscape architect. Swampy lands, foul water courses and steep slopes possess great possibilities for park development. A double service is thus rendered—primarily, the construction of parks—secondarily and incidentally, the elimination of municipal nuisances which are frequently of a very dangerous character. The foul and unsightly channel and its surrounding lands are transformed by the construction of sewers and the proper engi-

FIG. 1.—Swamp Land May be Utilized to Good Advantage in Park Development.

neering operations and subsequent planting into the mo attractive of parks, the channel itself providing what is ε especially pleasing and refreshing feature, a beautiful wat course.

The accompanying illustration shows a piece of swam land, almost worthless, which is now a part of tl Weequahic Reservation of the Essex County (N. J.) Pa Commission. The swamp has been dredged and the wat level has been raised. It is now a very picturesque lak surrounded by attractively planted uplands.

There are in almost every county spots of natural beau which it is the duty of the present generation to preserv The national government recognizes this principle in i great reservations, among which are the Yellowston Yosemite and Mount Rainier National Parks. One of tl most unpardonable losses in this connection was the cutti of the "Sherwood Forest" at Philadelphia, a tract forty acres of virgin timber actually within the city limit This land, which afforded natural advantages of very u usual merit for park purposes, passed into the hands of re estate operators and now is part of the two-story dwellin house district of West Philadelphia.

The Economic Advantage. The laying out of par increases the desirability of a town as a place of residenc Municipal boards of trade are realizing this principle increasing measure. Park development enhances the val of neighboring real estate. New parkways and boul vards afford very desirable frontages, the very best to l had for residential purposes. Much might be written the subject of real estate appreciation due to park develo ment; suffice it to mention here only a few illustrative case Real estate operators testify that the increased valuati in the vicinity of a properly regulated park will very so pay for the original investment. The assessed valuation in 1856 of the three wards adjoining Central Park, New York, was $20,500,000. In 1873 it had risen to $236,100,000. The natural increase as obtained by averaging the gain in the other wards was found to be $53,000,000, making the earning capacity of the park for the three adjoining wards, $183,000,000. In three years the valuation of the wards surrounding Prospect Park in Brooklyn rose $7,000,000, which was twice the cost of the land acquired for the park. The wisdom of the "Back Bay" improvement at Boston was certified in an increased valuation of surrounding property from 1877 to 1885 of $12,000,000 with a corresponding increase of revenue of $153,000. During this time new buildings were erected to the value of $10,000,000, giving a revenue of $128,000, making the total increased revenue, $281,000. In the city of Newark, property in the neighborhood of Branch Brook Park has increased in value to a marked extent, and to the Branch Brook development can be largely traced the great popularity of the Forest Hill residential section.

THE CLASSIFICATION OF PARKS

A convenient classification may be made thus: 1. National Parks; 2. State Parks; 3. County Parks; 4. City Parks; and 5. Parkways and Boulevards. The first two classes will not be treated in this discussion, since their character is principally that of great reservations of land of especial scenic and natural beauty. County Parks are usually more or less developed. They are suburban or rural in character and are usually of large area. City Parks include the old-fashioned city square and the modern city park whose development may be formal, informal, or both. When the governing board is a county park commission, city parks are classed as county parks. The Parkway and Boulevard are connecting arteries which join the parks of a system. The Boulevard is the more formal of the two and often is nothing more than a beautified avenue, while a Parkway is much broader, often about 400 feet wide, and may be laid out in a semi-informal manner.

Another classification used in some cities is, 1. City Parks; and 2. Outer Parks. The City Parks are those inside, and the Outer Parks those outside the city limits. The second division of this classification includes the reservation lands and the connecting parkways.

The accompanying map of the Essex County system, Fig. 2, page 11, shows a modern and scientific park development. But few of the original parkways as planned by Olmsted Brothers, the landscape architects, have been laid out, though some of these may come later.

ACTUAL RESULTS

The results of the last twenty-five years of this great movement have been especially remarkable, though a few of our older cities had parks prior to 1890. It is outside the scope of this book to make a full statement of park accomplishments in America, which are now very extensive, almost all of our larger cities having caught the spirit of the movement. In order, however, to give some idea of its development and magnitude, a few typical and important cases will be cited.

New York. The greatest pioneer work in America, though not the earliest, was the construction of Central Park. The land was purchased in 1856. The total acreage at present is 840 and the price paid for the land was $6,300,000. The total amount spent in bringing the park to its present condition is over $30,000,000. The park system now covers several thousand acres and includes Prospect Park in Brooklyn, one of our oldest parks, the Brooklyn Forest and Bronx Park with its Botanical and Zoological Gardens.

PHILADELPHIA. The original plan in Philadelphia consisted of five small squares to which Independence Square was afterward added.

Fairmount Park dates back to 1812, when it contained five acres. It was enlarged in 1855 to forty-one acres and in 1867 to its present proportions, mainly in the interests of the public water supply drawn from the Schuylkill River. The Park Commissioners were able to prevent the contamination of the river as far as their jurisdiction extended and to exercise a wholesome influence upon the Councils in the matter of proper sewer construction to intercept the flow from the river. A very interesting report made by a Special Committee of the Commissioners of Fairmount Park upon the Preservation of the Purity of the Water Supply, submitted October 11, 1867, and sent to the author through the courtesy of Mr. Carleton E. Davis, Chief of the Bureau of Water, and Mr. Thomas S. Martin, Secretary, contains full information in regard to this matter. In this connection it may be added that park lands may be located so as to combine the park needs with those of the water department. Several American cities have seen this advantage. Surface water supplies as impounded from small streams can be very satisfactorily guarded by the acquisition of the catchment area by the park department. This is one of the very best examples of municipal economy that can be mentioned.

The area of the Philadelphia parks is now over 5000 acres, 1000 acres of which is in Wissahickon Park, a beautiful ravine development, one of the best features of the Philadelphia system.

The development of the water fronts of the Schuylkill and Delaware Rivers is under contemplation as part of the general city plan.

The author here takes occasion to acknowledge the assistance of Mr. Andrew W. Crawford, Recording Secretary of the City Parks Association, who has furnished information as to the parks of Philadelphia and other American cities.

Chicago. The movement here dates back to 1869. One of the features of the park system is the incorporation of extensive recreation parks and playgrounds which have been secured in the past fifteen years.

Preliminary plans according to Mr. Walter Wright, Secretary, Special Park Commission, are now being prepared for the construction of a great outer parkway system which will give Chicago a continuous belt of parks and parkways around the entire city. This work has been authorized by the State Legislature subject to the approval of the voters of Chicago, which approval was received at an election in November, 1914.

Boston. This is a city of many suburbs, each separate in its local interests and government. These different municipalities have jealously guarded their local autonomy, but have wisely united to provide the general necessities, such as water, sewerage and parks. These necessities have been secured through the help of the State Government, which has created metropolitan commissioners with authority to provide them. Each municipality now has its local water supply fed from the metropolitan main lines; its local sewerage system, discharging into the metropolitan trunk sewers; and its local parks, secured either before or after the appointment of the Metropolitan Park Commission, which provide for local needs not met by the Metropolitan Parks. Boston has parks covering a total area of 500 acres; Cambridge has developed a frontage on the Charles River; and Lynn has acquired large tracts for park and water-supply purposes.

The Metropolitan Park Commission consisting of five unsalaried commissioners was created in 1892. Up to the present time it has purchased 10,250 acres of land, of which the largest tracts are the Blue Hills Reservation, twelve miles from the State House and easily reached by electric cars; and the Middlesex Fells Reservation, five miles from the State House. The metropolitan parks with their connecting parkways and seashore and river-bank reservations form a model system.

Essex County. With the exception of Military Park, a tract of but a few acres in the city of Newark and a few other small squares, Essex County had no parks up to the year 1895.

At a dinner in the city of Orange in January, 1894, a plan was suggested for obtaining parks and a meeting was soon after arranged and held in the rooms of the Board of Trade in Newark. Park committees from Newark and Orange were present and the plan previously suggested was approved. A committee was appointed to prepare a bill for the State Legislature which was promptly drafted and approved and then presented to the Senate at Trenton. It was passed and signed by the Governor early in May of the same year.

The bill authorized the presiding county judge to appoint a Commission of five persons to consider the advisability of laying out a system of parks and provided an appropriation of $10,000 to cover the salaries of assistants and

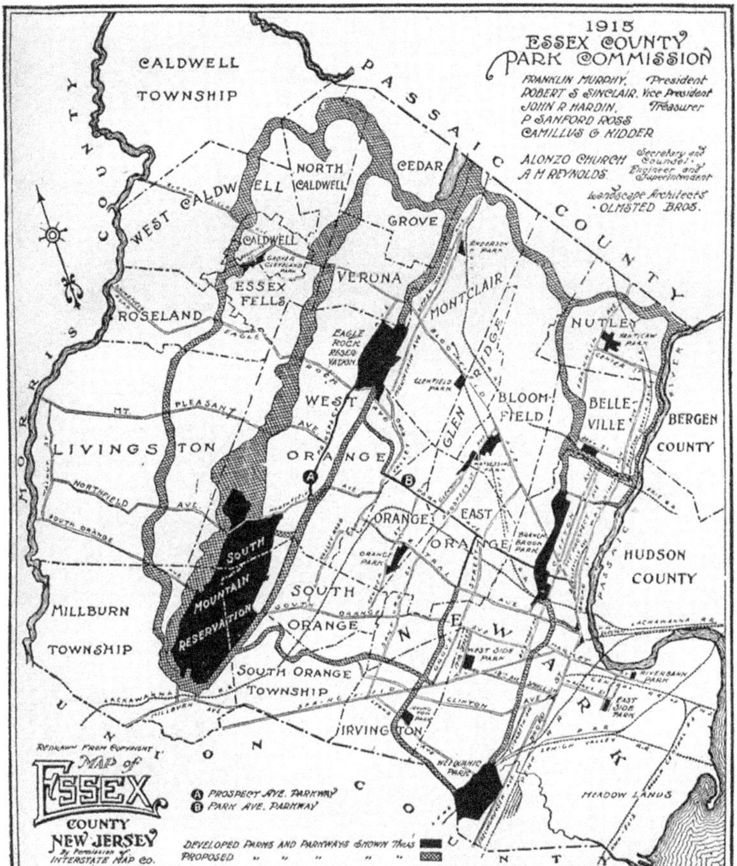

Courtesy Interstate Map Co., of Newark, N. J., engravers and publishers.

FIG. 2.—A Typical Park System.

traveling expenses of the members of the Commissioı The first Commission completed its work in 1895, aft‹ having reported favorably to the plan for parks, and su; gested a practical plan.

A new bill was passed by the Legislature in 1895 for tl creation of a second Commission to be appointed as befor This bill carried an appropriation of $2,500,000 of count funds and a referendum clause. At a special election tl bill soon became operative. The necessary moneys weı obtained by county bond issue by the Board of Freeholdeı and turned over to the Commission. Subsequent issues ‹ bonds were authorized, the outstanding indebtedness no amounting to $6,000,000.

There are now five principal parks in the system wit a total area of 3200 acres, most of which is mountai reservation.

THE FUTURE PROBABILITIES

The park movement is thus observed to be of recer origin. The United States in many respects has passe out from the period of mushroom growth into one of com parative stability. There are many evidences of thi: Flimsy bridges are being replaced by permanent masonr structures; wooden buildings by those made of fire-proc materials; cheap city pavements by those of a permaner character; railroads are being straightened, and elevate or depressed where they pass through large cities; railroa property line fences are being replaced by hedges after th English custom; and far-sighted business corporations aı making provision not merely for the present, but for severa decades to come. The policy of municipal wisdom call for the immediate setting apart of park lands to be pai cases will be followed. In some respects it is along the movement on foot for the conservation of our national resources.

THE WAY TO OBTAIN PARKS

The methods adopted have already been stated. A brief summary will be given.

1. Preliminary Agitation. Public sentiment frequently crystallizes in the appointment of a park committee or in a resolution passed by a board of trade.

2. The Creation of a Commission. If the work involves several municipalities, as it usually does, a bill for the creation of a park commission with delegated powers can be introduced into the State Legislature.

3. How the Necessary Funds are Provided. These are usually obtained from the sale of bonds, whose issue is authorized by the State Government, the indebtedness being incurred by the county or district benefited.

THE ORGANIZATION FOR WORK

The working organization is made up about as follows:

1. A *Counsel* and perhaps assistants to give legal advice, to attend to claims and to examine contracts and agreements.

2. A *Landscape Architect* to make selection of lands for parks, parkways, reservations and playgrounds. It is the function of the landscape architect to prepare all grading and planting plans and to outline for the Architect (see 5) the general features of buildings, gateways, walls, steps and bridges.

3. A *Purchasing Department* to obtain options on properties likely to be acquired and to arrange for the pur-

FIG. 3.—Submerged Land along the North River Later Filled in and Now a Part of Riverside Park.

chase of properties already embodied in the scheme of tl landscape architect.

4. An *Engineering Department* consisting of a chi engineer, assistant engineers, instrumentmen, chainme rodmen, inspectors and a chief gardener and his assistant The police department as long as the engineering depar ment is in existence is a part of it. When the work of lar purchase and heavy construction is completed, it pass over to the Superintendent's Department (see 6). Tl work of the engineering department can be divid thus:

a. Property surveys of lands to be acquired with tl accompanying searches in the office of the recorder of deed and the preparation of property descriptions to be inco porated in deeds of conveyance prepared by the couns or his assistants.

b. Topographical surveys prepared for the use of tl landscape architect in the preparation of his plans.

c. Designs of sewer, drainage, water-supply and lighti systems, together with the design of all the essential eng neering features of walls and bridges.

d. The preparation, letting and supervision of contracts for construction work. This includes the dire tion in detail of all construction operations.

e. Planting and gardening operations.

f. The maintenance of the work until it can be turn over to a Superintendent's Department, which need not organized until the work is well under way or even cor pleted.

g. The policing of all acquired areas. The police for consists of a chief and mounted and unmounted patrolm with sometimes a few plain-clothes men.

structures, such as buildings, gateways, steps and the architectural features of bridges.

6. A *Superintendent's Department,* whose function it is to take over the work done by the Engineering Department and carry it to completion. This work consists of gardening and planting operations, construction work (generally by day labor), and the ultimate maintenance of the entire system. The police department eventually is transferred to this department.

7. A *Clerical Force* to transact the work of the Secretary and Treasurer of the Commission, to issue permits for use of playgrounds, and to keep open office.

8. A *Title Guarantee Department.* It is customary to employ a title guarantee company to pass on all property transactions.

All of these departments may report directly to the commission, though there is some variation in this matter.

FIG. 4.—Reclaimed Land along the North River (the same location as shown in Fig. 3, page 15).

CHAPTER II

LANDS AND SURVEYS

GENERAL REQUIREMENTS IN THE SELECTION OF PARK LANI

THIS is specifically a question of landscape architectu It is also a matter of common sense. Many gre mistakes, though not chargeable to the profession of tl landscape architect, have been made in the selection park lands. The following are the leading requirements

1. Parks must be accessible. City parks should l so placed that all of the population can reach at least o park on foot. The larger county parks should be easi reached by carriage, automobile and trolley car. Tl fare on these car lines should be reasonable. Excessi charges should not be permitted.

2. It seems to be the modern idea that a park syste should be a connected system, the various parks bei joined by parkways or boulevards. This is regarded as very essential matter.

3. A park should be located with regard to the oth requirements of the territory in which it is placed. Centr Park in New York City, although an asset of inestimab value, is so placed that it cuts off two of the main arteri of travel running north and south on Manhattan Island.

4. Parks should be planned with a distinct view of tl requirements of the population to be benefited. Tl problem must be worked out for each individual case. Tl needs of the entire population must be considered wi

regard to wealth, culture, nationality, age, sex, density of population, etc. Drives, parkways and boulevards will principally benefit the wealthy. A fine landscape, beautiful architecture and opportunities for nature study meet the cultural requirements. The inherited traits of people of different national extraction should be regarded. The matter of age must also be carefully considered; while the older person will generally expect opportunities for quiet and rest, children and young people will need playgrounds and athletic fields. It may here be said that the playground movement has recently assumed enormous proportions as one of the best opportunities for civic improvement. The child should have his swing, sand box and wading pool; the young woman her tennis court, and the young man his baseball field and out-of-door gymnasium. Where the population is dense, the park would seem to be the only solution of the pressing problem of rest, recreation and amusement.

5. It is a fact well known to experts that lands which are poor from the point of view of the real estate operator are often the best for park development. Low and swampy lands which are usually festering spots in cities are among this class. A very good illustration of what is meant by this paragraph is found at Philadelphia, where portions of the Tacony, Pennypack and Cobbs Creek valleys are now on the city plan to be used for park purposes. These valleys are narrow and deep. If laid out in city blocks, the grade of the cross streets would be about that of the sides of the valleys. An enormous amount of filling would thus be required, the cost of which would have to be borne by the city. The cost of this filling alone would be more than the purchase price of the tracts for park purposes.

THE ACQUISITION OF PROPERTY

The Survey. Before lands are taken over for park purposes, a survey is usually made by the engineering department. This is an easy matter when the boundaries are well defined by street lines, fences, stone monuments, hedges, stone walls or a sufficient number of blazed trees. It frequently happens, however, especially in rural reservations, that the landmarks, usually blazed trees, are partially or entirely gone. In such cases the determination of the boundaries becomes a difficult or impossible matter, and especially, as frequently happens, if the recorded deeds contain inaccurate or grossly incorrect descriptions. In all such cases the matter has to be adjusted between buyer and seller, and, if no agreement can be reached, the question is passed upon by a condemnation commission, to be described later on. In purchasing a large reservation, it is the usual practice for the engineering department to prepare a large map of the tract, locating the boundaries and the principal topographical features, such as brooks, buildings, etc. The surveys of the individual lots and, when surveys cannot be obtained, mapped deed descriptions are plotted on drawing paper or cardboard. These plots are then cut out and fitted together on the large map. Overlaps and underlaps are frequent in cheap rural lands and occasionally a triangular piece of land will have to be purchased twice.

Options. The policy of obtaining options on property likely to be acquired is a good one. Property owners can frequently be approached when they are in a favorable mood and advantageous agreements for the sale obtained subject to a definite time limit of purchase. These options may be obtained by members of the engineering force or by the purchasing agents.

Direct Purchase. Property to be taken over for park purposes on which no option has been obtained can almost always be purchased at a fair market price, it being to the advantage usually of the owners to sell, since, if they refuse, the matter can be taken before a condemnation commission and a forced sale brought about. The cost of litigation is sometimes great and the part borne by the owner frequently a total loss. Purchases are made through the purchasing agents.

Condemnations. The resort to this method of purchase is somewhat rare, since, as before stated, it is almost always to the advantage of the owner to sell. Condemnation proceedings for park lands are carried on about as follows. The method given is for the State of New Jersey:

1. A notice to one of the justices of the Supreme Court that agreement cannot be reached for the purchase of the property, together with an application to said justice for the appointment of a condemnation commission of three men to examine the land and make a just appraisement. This notice is endorsed by the justice, who assigns a time and place for the appointment of the commissioners.

2. A notice from the park commission to the owner and parties interested of the previous application for the appointment of condemnation commissioners, together with a copy containing the affidavit of the person who served the notice on the owner and parties interested, which copy is kept on file in the office of the Commission.

3. A notice from the Supreme Court justice to the park commission of the appointment of a condemnation commission together with directions in regard to the notification of all parties concerned in the property sought to be acquired. This document also includes the oaths, before a person duly authorized to administer them, of the

condemnation commissioners faithfully to appraise the lands.

4. A notice from the park commission through its counsel to the parties concerned in the transfer of the property, of the appointment of the condemnation commission and of the time and place of meeting to view and examine the lands and rights therein in order to make an equitable appraisement. This document contains the oath of the notice server.

5. A report of the condemnation commissioners as to the value in a gross sum, of all the interests, estates, or shares in said lands, whether in possession, remainder, reversion or expectancy.

6. The certificate of the title guarantee company as to the soundness of the title.

7. The deed.

THE TOPOGRAPHICAL SURVEY

Before preparing the designs, full information must be obtained of all topographical features. The method of making the survey will depend entirely on the character of the land to be surveyed and also on the probable mode of treatment. If the tract is to be entirely regraded it will be well to make a very accurate survey, which can also be used later on in determining the amount of excavation. Thus one survey may be made to answer two purposes and much time and money be saved. The method of making the survey is a matter of judgment and no hard and fast rules can be laid down. The subject will be briefly treated under three headings: Parks, Reservations and Parkways.

For Parks. The survey for city parks, and frequently for reservations also, is made by the gridiron method.

The first thing to be done is to locate and properly

monument a suitable base-line. This base-line may be placed on one side of the tract if it has a long straight side. Hubs 3 inches square are driven every hundred feet and line and distance are marked by a tack or small nail. It is very desirable to locate this base-line where there is to be no cut or fill. Otherwise it will be disturbed and

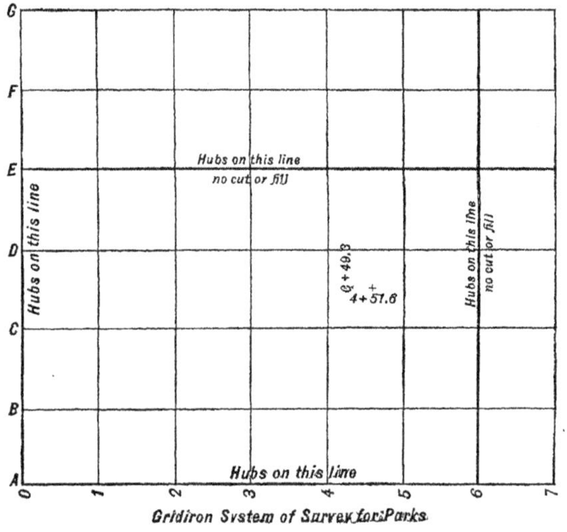

Fig. 5.

will have to be relocated. The importance of this remark cannot be too strongly emphasized. An auxiliary line of hubs is laid off at right angles from the first one. The lines at right angles to the first line of hubs may be designated by letters, thus—A, B, C, etc. The lines at right angles to the second line of hubs may be designated by numbers—0, 1, 2, 3, etc. Thus the corner of any square in the checkerboard layout can be indicated; as, C–4, which is a corner 400 feet from the main base-line and 200 feet from the auxiliary base-line. Also any point whatever on the tract can be designated, as, C+49.3–4+51.6. If the tract is a square one, it will frequently be well to have monumented cross-section lines on the four sides. If of irregular shape, well-monumented lines may be located at various distances apart, such as 1000 feet, where they are not apt to be disturbed or covered up by excavation or embankment.

In addition to the permanent cross-section lines, all other cross-section lines are also staked out by stakes about 1½ or 2 inches square. These stakes are allowed to project from the ground 6 or 8 inches and are marked with lumberman's chalk. The letter designation is marked on one face and the number designation on a face at right angles to it. The markings are so placed that if a person walked from the beginning of a cross-section line the distances out would be visible on the face of the stake directly in front of him.

After the cross-section lines are all staked out, the level parties begin their operations, which consist in obtaining the elevation of the ground at every stake to the nearest tenth of a foot. All changes in grade should also be determined. All topographical features must be located, such as streets, buildings, water courses and trees; also all sewers, water pipes, gas pipes and wire conduits. In locating the trees, each tree should be tagged with a copper tag ½ inch by 1 inch with a number stamped upon it. The diameter, kind and spread of each tree should also be determined. This information is very important to the landscape architect. If the trees are very close together, as they frequently are in the woods, instead of attempting to plot each tree in the note-book, a table may be prepared with the following columns.

FIG. 6.

Kind.	Diameter.	Spread.	Co-ordinates.	
			Letter.	Number.

The notes are kept in a cross-section book about $6\frac{1}{2}$ by $8\frac{1}{2}$ inches, ruled in blue lines, ten to the inch. In work of this kind it is especially important that the date should be entered on each page.

A map is then prepared on a suitable scale, depending on the amount of detail to be plotted. The following scales are used; $1''=30'$, $1''=40'$, $1''=50'$ and $1''=100'$. This map is traced and the tracing forwarded to the landscape architect.

For Large Reservations. The previous method should be used if possible. Since, however, no grading is usually contemplated except for wood roads, etc., and since it is often impossible to run straight lines, the method of survey by random lines, as nearly straight as possible, is adopted. Side shots are taken by angle and stadia. This method is not recommended when trees and other details must be located.

For Parkways. The method here to be followed is that of carefully running a transit line through the middle of the ribbon-like area, placing stakes every hundred feet.

Courtesy Alonzo Church, Secretary, Essex Co. Park Commission.

FIG. 7.—Topsoil Stripping, Piling, and Grading.

All angles should be "doubled" and checked by the needl and all distances should be checked by stadia. Si shots are obtained by angles and stadia distances. Th method of survey is not as accurate as the gridiron metho but the latter in its simplest form is not adapted to par

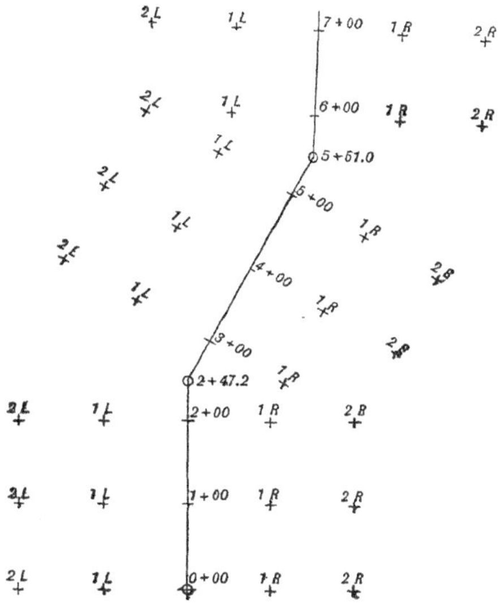

Modified Gridiron System of Survey for Parkways.

FIG. 8.

way work. A modified girdiron system, however, ma be used when the base-line instead of being one straigl line is the broken line of traverse.

General Remark. In making topographical surve it is sometimes advantageous to sketch in field. Work this kind is usually much more accurate, especially wh the map must be drawn in minute detail. It is advantageous where there are terraces, trees, etc. Mistakes in plotting are thus avoided, since the features are within sight of the draftsman. A plane table is a useful instrument for this kind of work; but it can be done with a transit with vertical circle, the draftsman using a light table 2 feet square mounted on a light tripod.

THE HYDROGRAPHICAL SURVEY

Ponds and Lakes. The topography of the beds of ponds and lakes is obtained by the gridiron method of survey. The depths may be had in shallow ponds by use of a rod lowered from a boat, the readings being referred in measurement to the water level. When the depths are more than 6 feet, a sounding line is employed. A good sounding line can be made of sash-cord with a colored string tied at each foot-mark. Different colors are used to good advantage, the number of feet being thus readily determined. A sinker will be required to hold the lower end of the line to the bottom. Alignment of the sounding rod or rope may be had by means of rods on the shore or else by transit. The distance is obtained by stadia or else by alignment rods on a line at right angles to the other.

Brooks and Creeks. The flow of streams is determined by use of a weir unless the flow is large. In the latter case, which is not common, a current meter or rod float must be employed. Fig. 9 shows clearly the customary method of making weir measurements.

Depths over the crest are measured either by a carpenter's square or a footrule. For measuring depths the hook-gauge is never necessary except when litigation is probable. Measurements of flow may be made twice a day, or oftener when the water is rising or falling rapidly.

Loaned by J. & W. Jolly, Inc., Holyoke, Mass.

FIG. 9.—Gauging the Discharge of a Stream.

In determining the discharge of a stream by mea
of a weir great importance should be attached to the matt
of keeping the weir crest free from floating objects, whi
by backing up the stream may invalidate the readin
even after the brush is removed. Weir frames are oft
washed out by freshets. To prevent this the frame shou
be firmly set in a trench excavated for the purpose wi
riprap on the down-stream side to prevent scour. Shou
the weir be found leaky or should leaks be apprehended,
little clay placed on the bed of the stream against t
upper face will be found very satisfactory.

Full hydrographical information is required by the lan
scape architect in his preparation of plans for lakes a
water courses, and by the engineer in connection wi
excavation and dredging operations and the design
outlet sewers.

CHAPTER III

DESIGN

The Architectural Treatment

THIS portion of the design is the function of the Landscape Architect. It is embodied in the following drawings:

1. A general grading plan usually drawn and traced on the scale of the topographical survey. By superimposing one tracing on the other, a composite print can be made showing the cut or fill at every point.
2. Cross-sections of drives, paths, terraces, etc.
3. Various plans and suggestions as to masonry work, bridges, etc.
4. Planting plans.

The Requirements. This book is not the place for a discussion of the principles of landscape architecture. The work is done by experts in this line of work. A few of the general principles are here suggested.

First of all, the plan must be a suitable one. An English garden would be inappropriate in a mountain park and the informal treatment would be inappropriate for a city square.

Second, the general principle of harmony must be observed. For instance, a cut-stone arch would be out of place in a natural ravine with wooded slopes.

Third, the design should be suggestive. For example, the idea of steepness can be brought out by planting trees on the crest of the slope.

Beautiful views should be preserved wherever possible

Courtesy Alonzo Church, Secretary, Essex County Park Commission.

FIG. 10.—Topsoil Pile.

Courtesy Austin-Western Road Machinery Co.

FIG. 11.—Machine for Stripping Topsoil and Grading.

and should not be shut off by trees. Vistas are design by a careful inspection of the topography and timb growth. Free access should be had to all parts of tl park. This does not mean that access is to be obtain necessarily by the construction of straight drives aı paths. Straight lines should not be used except in form architecture, where they may be entirely appropriat Many unsightly objects outside the park boundaries cı be hidden from the inside by the construction of bord mounds with suitable planting. These mounds are ve effective and are extensively used.

The Engineering Design

UNDERDRAINAGE

Wherever soils are swampy or wet, underdrainage necessary. This can be accomplished in two different way

First, by the laying of agricultural tile. Tile is ma in two patterns—horseshoe and cylindrical.

Horseshoe tile may be laid on a line of 1-inch boards else on the bottom of the trench if it is carefully grade The tile are frequently 2 feet in length. No coupling used, but a piece of cheese-cloth may be employed to goc advantage in wrapping the joints to prevent the entrance earth.

Cylindrical tile come in 2-foot lengths and are provide with 4-inch collars or rings to be used at the joints. Tl sizes commonly used are 2-inch, 3-inch, and 4-inch. Tl tile are porous and should be straight, hard and of unifor cross-section. In draining lands they are laid in parall ditches joining the main at 45 degrees. Y-connectio can be purchased for the purpose. The distance apa of the trenches, their depth and the size of the tile a matters which experience only can satisfactorily decic

The determining conditions are, the kind of soil, slope of the land and elevation of the water-table. The thoroughness of drainage is another feature. Waring's rule is frequently used in the design.

The mains are of porous tile up to 4 inches in diameter and above that of vitrified sewer pipe. If the line discharges into an open channel, the detail at the outlet is

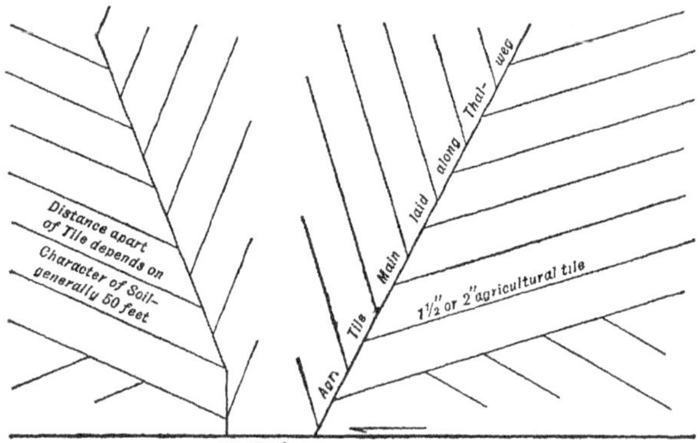

Vitrified Sewer
Method of Providing Underdrainage for Damp or Swampy Soil. Surface Water should be Removed by Inlets and Catch Basins along Upper Sides of Paths and Drives.

FIG. 12.

very important. To prevent injury and dislocation to the line, the use of a 12-foot length of cast-iron pipe is recommended. This pipe does not have to be perfect. A piece of " seconds " can be used to good advantage.

Tile drains become clogged in a few years, especially when in the neighborhood of willows and other trees which absorb much water. But the drainage process still continues in some measure, due to increased porosity of the

Fig. 13.—The New Bay Ridge Parkway. Surface Water Carried in Paved Gutters.

soil by the dissolving out of soluble matters after tl laying of the tile. Furthermore, the land will never b come as wet as it originally was on account of the fact th: the surface water is rapidly disposed of by the surface-wat drainage system.

Secondly, land can be subdrained by the construction stone or brush ditches. The principle of action is the san as in the case of the tile ditches. Stone or poles are placε in the bottom of the trench to provide open space for tl passage of the water. To prevent clogging up with eart the stone or poles are covered with branches or straw, c which the backfilling is deposited.

General Remarks. Trenches for underdrainage a usually spaced 50 feet apart. The depth varies from 1 feet to 4 or 5 feet. The slopes should not be less tha say 0.25 of a foot in 100 feet except for the mains.

SEWERS

Park sewers are of three kinds: 1. Storm-water Sewer 2. Sanitary Sewers; and 3. Combined Sewers. The metho of design and construction are in some respects differe from those of municipal sewers.

Sewers are usually laid along straight lines as in m nicipal work, but the great expense of manholes can l saved in whole or in part. Park sewers seldom lie und pavements. If they become stopped, the trench can l reopened and the obstruction removed. It will be cheap to do this occasionally (although it hardly ever becom necessary) than to go to the great expense of placing ma holes at every change of line and grade.

Park sewers are of brick or vitrified pipe. Sewers larg than 36 inches in diameter are built of brick. When tl soil is treacherous, steel or spiral-riveted pipe may som times be used to advantage where dislocations might otherwise occur.

1. Storm-water Sewers. The peculiar features of park storm-water sewers are the methods of collecting the water, the methods of discharging it and the calculation of size.

Methods of Collection. In order to collect storm water, inlets and catch-basins must be constructed.

The inlet is merely an opening for the admission of water. It consists of an iron grating set in concrete over a 6-inch elbow from which the water is carried by a 6-inch chute into a catch-basin or directly into the sewer.

The catch-basin is usually built of brick, of circular horizontal section and surmounted by a grating to provide for the entrance of rain water and the exclusion of sticks and leaves. It is constructed about 6 or 7 feet deep, 4 feet inside diameter, corbeled in at the top to a diameter of 2 feet. The cast-iron grating is rectangular for paved gutters and circular for grass gutters. Its height is about 8 inches. It is cast in two pieces—the lower part provided with a flange to rest upon—and a grating to hold back obstructions. A catch-basin is provided with a concrete foundation about 6 inches thick. The invert of the outlet pipe, which is 6 inches in diameter, is about 3 or 4 feet from the bottom. No outlet trap is provided, the purpose of the catch-basin being merely to catch mud and gravel and prevent its entrance into the sewer.

Instead of placing two catch-basins on opposite sides of a path or drive, it is often well to dispense with one and build an inlet instead. The inlet discharges into the catch-basin and the catch-basin into the sewer. Should the inlet or its chute become stopped, it may be cleaned out by use of a garden hose.

Catch-basins and inlets can be placed in paved gutters,

Courtesy Alonzo Church, Secretary Essex Co. Park Commission.

Fig. 14.—Park Walk along Steep Terrace Showing Paved Gutters.

or off to one side of the path or drive in grass gutters, which will be discussed under the heading of Grading. The distance apart is a function of the amount of water to be carried. Past experience has clearly demonstrated that catch-basins and inlets are frequently placed too far apart and almost never too close together. A spacing often used is 125 feet.

If a sewer carries water from a row of catch-basins along a drive, the sewer should not be placed under the drive, but rather under grass or shrubbery, where it can easily be reached in case of trouble.

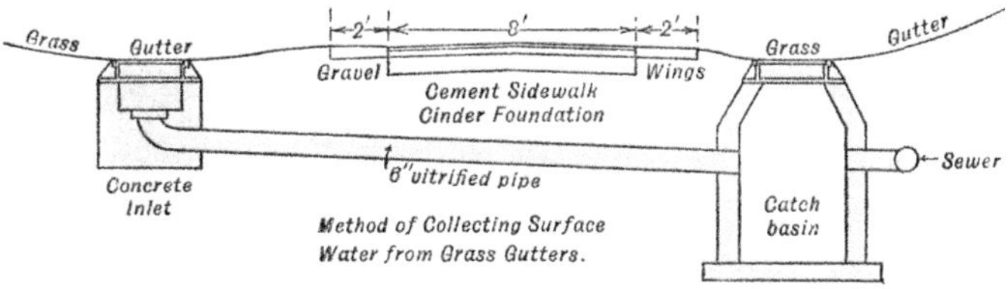

FIG. 15.

Methods of Discharge. If storm-water sewers discharge into other park sewers, it is not necessary to construct a manhole, as is done in city practice, providing that due care is observed in making the connection.

If a storm-water sewer discharges into an open body of water, the discharge may take place below the surface of the water. This means that the outlet must be of special design. Cast-iron pipe held in place at the end by slope paving is sometimes used. Or in other cases the sewer may pass out through a shore wall which holds it in position.

The turbidity of the first wash is often a source of trouble by reason of the fact that the pond or lake is rendered turbid. The first washings may be side-tracked into an outlet sewer through a pipe too small to carry the storm-water flow.

Sometimes where there are springs or another adequate source of supply for the lake or pond, the entire volume of storm-water is passed off into the outlet sewer direct.

Calculation of Size. The usual methods of calculation are employed. The coefficient of imperviousness, however, will be low, due to the character of the surfaces, which are mostly grassy.

An outlet sewer from a lake need not be of sufficient capacity to carry the entire flow of the storm if this flow is admitted to the lake. The body of water may be used as a reservoir whose elevation will rise slightly at time of a storm and the surplus water will run off through the small outlet sewer after the storm is over.

2. Sanitary Sewers. These are used to carry the flow from toilet rooms which are located in shelters and pavilions. Such sewers are small. They should never discharge into park lakes and water courses. Sanitary sewers are generally laid much deeper than storm-water sewers. It is very important that the grade be flat in order that the depth of flow may be as great as possible. The minimum velocity should be 1.5 feet per second. In order to lay the sewers on flat grades it may be necessary to build a few drop manholes at places where the sewers join. The size of sanitary sewers is 6 inches except for the mains. Owing to the impossibility in many cases of obtaining a proper depth of flow, say one-quarter of the diameter, flush tanks will have to be used to keep the sewer free from incrustation.

3. Combined Sewers. These are but seldom used in park engineering. Occasionally however, where a sanitary sewer is available, storm-water sewers will be connected

Courtesy Buffalo Steam Roller Co.

FIG. 16.—Park Walk. Drainage in Grass Gutters. Rolling being Done by Special Tandem Roller.

to it, thus making a combined sewer. This means enlargement of size. Difficulty is often experienced maintaining in dry weather a proper depth of flow with(flushing. The flow from these sewers should not be c charged into lakes and water courses unless there is a p vision for removing the dry-weather flow to another outl

GRADING

The grading plan is prepared after a careful examinat of the topographical map and a personal examination the site. This is the function of the landscape archite Some of the features in the design directly affect the dra age and other plans and will thus be explained.

Grass Gutters. Instead of allowing the sides of dri and paths to act as channels or gutters for the collection a removal of water, it is customary in modern practice provide grass gutters wherever possible. These cons of depressions 4 to 6 inches deep and 6 to 12 feet wide the sides of drives and paths, running parallel with the (See Fig. 15, on page 47.) In vertical cross-section t traces of these gutters should be curves which are co tinuous with the lines cut from the adjacent slope. Th gutters possess marked advantages. They intercept stor water and prevent it from washing away the surfaces drives and paths. Since the flow of water through them gentle by reason of obstruction by blades of grass, there practically no erosion and very little solid matter is deposit in the catch-basins. In fact in designing the draina system some engineers dispense with basins altogether a merely use inlets. On the lower sides of driveways a paths, the gutters of course are unnecessary. There a times, however, when by reason of the steepness of t gutters cannot be used. In such cases it will be necessary to pave the gutter so that it may carry the run-off from the slope and drive or path.

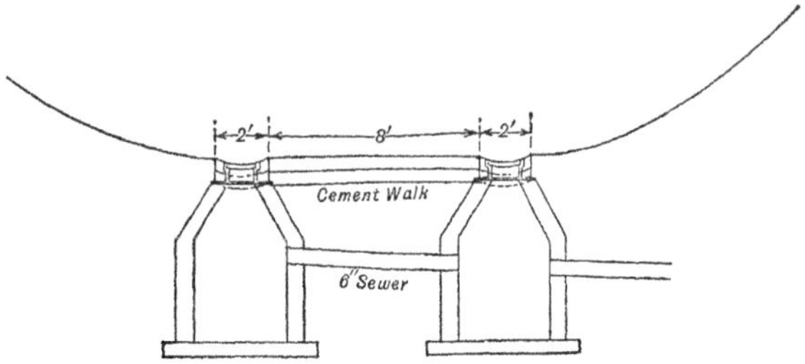

Method of collecting surface water from paved gutters which are necessitated by reason of steep side slopes. Inlets not satisfactory on account of large amounts of grit.

FIG. 17.

Figure 13, on page 41, furnished by Mr. Cabot Ward, Park Commissioner of the Boroughs of Manhattan and Richmond, New York, illustrates the proper treatment for drives when the side slopes are steep. Paved gutters here carry the surface water.

Absolutely level fields cannot be drained. Slight depressions or slopes should be provided so that surface water may be promptly removed. If the work is carefully staked out, these slopes and depressions need be but slight.

Lakes and ponds should not drop off abruptly at the shore line. It is safe and proper to provide a shelf at the shore line extending out 8 to 10 feet with a drop of $1\frac{1}{2}$ feet. Lakes and ponds should be at least 8 feet deep if possible. If shallower, there are apt to be objectionable growths of water plants. (See Fig. 18, on page 55.)

PILING AND BULKHEADS

Pile Foundations. These are often necessary in quicksands and muck to support bridges, overflows, etc. Two kinds of piles are used, timber and concrete. Timber piles will answer the purpose if not too expensive and if they are driven below the water table. The soil surroundiny them may, however, afterwards become dry due to the construction not far away of deep brick sewers. If the piles dry out they are sure to rot. Wooden piles are driven by the familiar "land machine," with hammer weighing from 1500 to 2000 pounds, or else by the steam hammer. The latter is not so satisfactory in peaty soils, due to the presence of snags, which can often be shoved aside by the vigorous blows of the drop-hammer machine. Wooden piles are sometimes jetted into place by the hydraulic process which, it is almost needless to say, can be used only in sandy soils free from stones. Instead of sawing off and capping driven piles, it is better to make a foundation of concrete. This is much cheaper and is also stronger, since the concrete surrounds the piles and holds them rigidly in position.

Concrete piles are also used and may be of the type that are driven or jetted into position or else of the type that are cast in a hole prepared by the driving of a metal form. Concrete piles have the advantage of durability, and when reinforced a marked advantage as columnar piles. They are thus especially adapted to semi-fluid peaty soils.

The safe load to be borne by a pile is a difficult matter to determine. Arbitrary rules have been employed and when used by experienced engineers give results on the safe side. Considerable literature has recently been published on the subject of piles and pile driving, for which see Transactions of the American Society of Civil Engineers, Vols. XLVIII, p. 180; LIV, Part F, p. 3; LXV, p. 467; LXX, p. 412.

Bulkheads. In constructing artificial lakes, it sometimes happens that a part of the shore lies in an unstable soil which cannot be excavated without flowing. To construct a firm and stable shore the following method has been used:

A single row of piles close together with every fifth one omitted is driven along the shore line. Then batter piles are driven in the gaps at as great an angle with the vertical as possible. Unfortunately with a drop-hammer machine this angle cannot be made greater than about 20 degrees. The piles are then all sawed off at about 1½ or 2 feet below the proposed water surface and a wale is framed into them in such a way that the thrust of the batter pile due to lateral movement of the soil is transmitted to the wale and then against the vertical piles. (See Fig. 18, page 55.)

In performing this work of cutting off the piles and placing the wale in position, a trench must be excavated in order to provide room for the workmen. In order to draw the work together jack-screws are used and a bolt with suitable washers through each pile holds it close to the wale. The wale consists of an 8-inch by 10-inch timber. The piles are notched to receive it, since the bolts must not be required to take up much of the thrust. The material on the water side of the piles can then be removed by cable-way or otherwise to the full depth of the lake. Good upland subsoil is then dumped in over the piles to reinforce them and provide a firm shore. This soil of course will sink down into the soft material below and more will have to be added until no further settlement takes place. It is wise to fill in considerably above the proposed grade

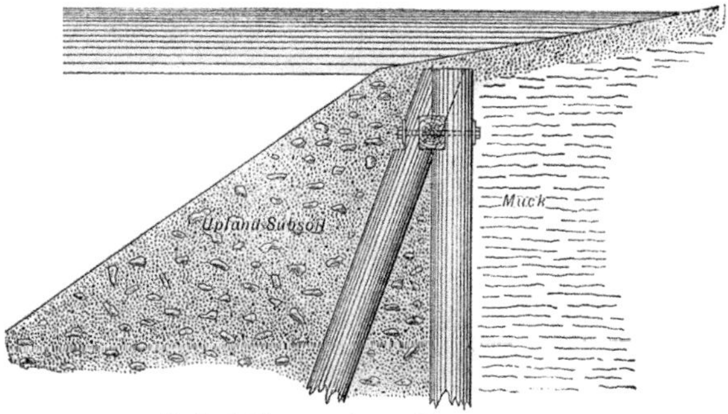

Timber Bulkhead and Subsoil Filling to Form Artificial Shore of Lake Excavated in Poor Soil.

FIG. 18.

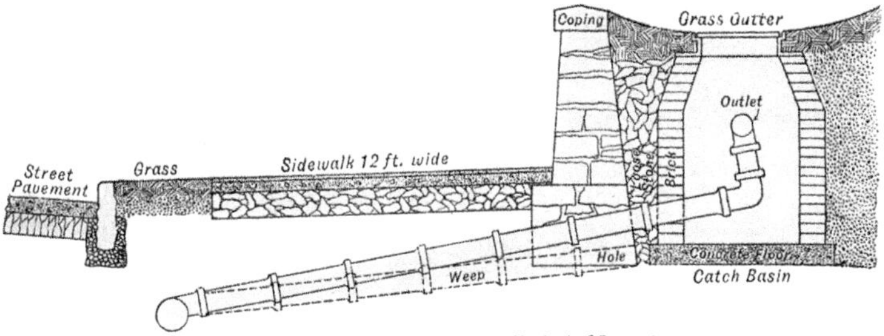

Method of Removing Ground and Surface Water Collecting behind Retaining Wall.

FIG. 19.

in order to ensure stable equilibrium and then, after s a month, remove the surplus material. As the good s sinks into the soft material in which the lake is being co structed an upheaval of course takes place, which necess tates further excavation.

In order to provide a solid bottom for the lake, 1-in boards are laid loosely on scantling and covered with inches of good subsoil.

MASONRY WALLS AND STEPS

Retaining Walls. It is essential that masonry wall whether of stone or concrete, be kept dry. In order secure this, proper provision must be made both for tl collection and for the disposal of water.

If a retaining wall is backed by a grassy slope, a gra gutter should be provided to intercept surface water befo it reaches the wall. Catch-basins or inlets are placed this gutter at proper intervals, usually about 125 feet, ar the water entering them is carried off through a storn water sewer. (See Fig. 19, page 55.)

Ground water must also be taken care of. To do th it is well to place broken stone behind the wall to act as reservoir and passageway for the water, which can l drawn off through the wall by means of " weep holes spaced from 15 to 50 feet apart. A square hole in the wa about 4 inches by 4 inches in cross-section answers the pu pose. Or a 3- or 4-inch iron pipe of the kind commonl called " seconds " may be used instead. The first methc is frequently followed in stone retaining walls and tl second in those made of concrete. The weep hole ma discharge directly upon the surface, or the water may l carried off below the surface. The first method will suffi if no path or walk is adjacent, since the quantity of wat discharged is slight. But if there is a walk near the wall, it is better to follow the second course. The flow from the weep holes may be carried below the walk in pipes and discharged into a street gutter through holes in a concrete curb, or the pipes may discharge into a line of agricultural tile or vitrified pipe.

Perfectly straight and continuous walls running parallel to a street are sometimes extremely monotonous. Variety can easily be secured by the use of pilasters or buttresses.

The customary rules for thickness cannot be followed in designing low stone walls, say 3 or 4 feet high above the foundation, unless the stone have true horizontal beds and extend entirely through the wall. Since this is seldom the case, the wall should have a thickness on top of the foundation of at least one-third of the height. The thickness under the coping should not be less than 15 inches.

The foundation offset should never be visible, but should be covered with a few inches of soil.

Dry walls are frequently advantageous by reason of their cheapness. Stone uncovered by the grading operations, if of the flat variety, such as shale rock, is excellently adapted for the purpose.

Steps. These may be of stone, usually granite, or of concrete. They should be designed with broad treads and low risers. The treads are about 14 inches wide and risers about 6 inches high. The exposed corners should be slightly chamfered.

If a flight of steps is placed in a curved walk, the steps should run truly normal to the curve, the walls at the sides being either continuous with the path lines or else tangent to these lines. Stone steps should be so designed that they rest for several inches one upon another, and all vertical joints should be broken.

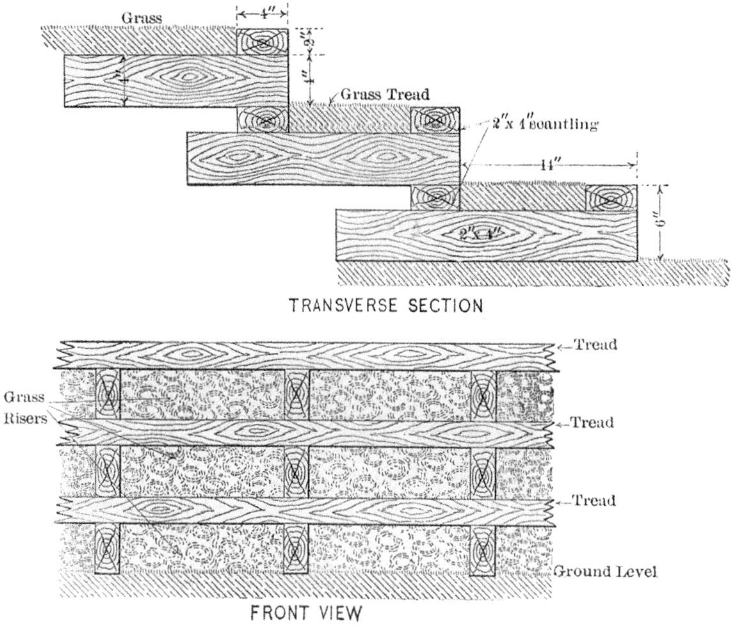

Fig. 20.—Sod Steps.

Cement steps should be finished rough with crushe quartz or coarse sharp sand.

A very attractive flight of sod steps was shown to tl author by Mr. G. A. Parker, Superintendent of Parks, : Hartford, Conn. To prevent the breaking down of tl edges, a scantling was placed at the face of the tread ar supported as shown in the illustration, Fig. 20, page 59.

The grading above a flight of steps should be of such character that no water will reach the steps. This end best attained by use of grass gutters.

WATER PIPES AND FIXTURES

All pipes above 3 inches should be of cast iron. Tl smaller sizes, 1½-inch, 2-inch and 3-inch, are of galvaniz wrought iron. Most of the pipes in a park system a of the smaller sizes. Much expense can be saved l laying the pipes in shallow ditches, say 3 feet deep. Th can be done only when no water is needed in the wint time. If pipes are laid in shallow trenches it is imperati that they be laid to grade, and that all low points be pr vided with blow-offs into an adjacent sewer. If no sew is low enough, part of the system will have to be laid at depth of at least 5 feet in places whose climate is like th of New York City.

Water pipes supplying buildings must always be la at a sufficient depth to prevent freezing in the winter time.

Dead ends should be avoided as much as possible. Tl usual gridiron system should be followed.

The hydrants used, with the exception of a few fi hydrants required, are of the lawn pattern. The ordina lawn hydrant does not project above the ground. It provided with hose connection and is opened by a sho key about 3 inches in length. Occasionally water cran are used for filling water wagons. These should be simple and of neat appearance. The hydrant shown in the figure is provided with a valve key for shutting off the water below ground and emptying the riser. If hydrants are set in swampy ground, special provision should be made to take care of the drip. This can best be removed by making a sewer connection. (See Figs. 21 and 22.)

The water supplied to parks is frequently metered. For small consumption, a water meter of the disc, rotary or reciprocating variety is used, but for large consumption a Venturi meter of the self-recording type is needed. This meter requires the erection of a small house to contain it. Water meters should never be set in wet manholes. If the manhole is below the water table, the soil should be drained or else a drain pipe should extend to the bottom of the manhole. If the soil is usually dry but occasionally wet, the pit can be kept dry by building it of a ring of brickwork $1\frac{1}{2}$ bricks thick and thoroughly filling the joints with impervious mortar.

Drinking fountains are placed where needed. The water from galvanized pipes has an unpleasant taste when the pipes are new, but the disagreeable taste gradually disappears.

PATHS

Paths are surfaced with cement, with gravel and occasionally with asphalt or brick. Before considering in detail these different kinds of walks, a few remarks will be made in regard to (1) the provision for drainage, (2) the shape in plan, (3) wings, and (4) entrances.

(1) *Provision for Drainage.* Except in dry, gravelly soils it is customary to lay cinder or broken stone foundations to provide drainage and prevent heaving from frost. All walks should be on an incline. At low points provision

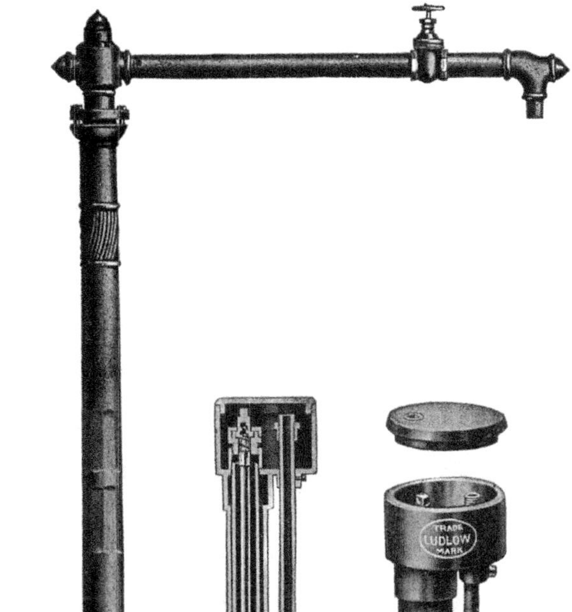

Courtesy Ludlow Valve Mfg. Co.

Fig. 22. Fig. 21.

Water Crane and Lawn Hydrant.

should be made for the removal of the water whicł collected. This can be done by providing a chute of cultural tile to the nearest catch-basin or sewer. D age for street sidewalks can be had by running a short to the gutter.

If a path is close to a steep terrace, it may be nece: to carry surface water in a gutter in the path itself. 14, page 45.) In the case of gravel walks or of cer asphalt or brick walks with gravel wings, the gutter be made of small stone blocks. Inlets and catch-b will have their gratings in the gutters. These gra should be slightly depressed or else the water will them without entering. If the slope is steep the gra should be placed level or sometimes at a slight inclin: against the grade. If the walk is nearly level it ma; be necessary to pave the entire gutter. It is often en to place the blocks every foot or two, which is sufficie prevent the washing away of the gravel.

(2) *Shape in Plan.* Park walks may be either str: or curved, depending on the nature of the design. the formal garden the straight walk is proper, but i informal design straight lines should be avoided. C should not be laid out arbitrarily. A curve should al have its reason. This is afforded by a natural obsta by the topography. If natural obstacles do not exist, can be introduced in the design.

(3) *Wings.* In order to increase the effective wid a park walk, wings of gravel 4 feet wide on each sid often used. These wings may have the full cinder dation as in the cement, asphalt or brick pavement. economical construction the cinder foundation is omi

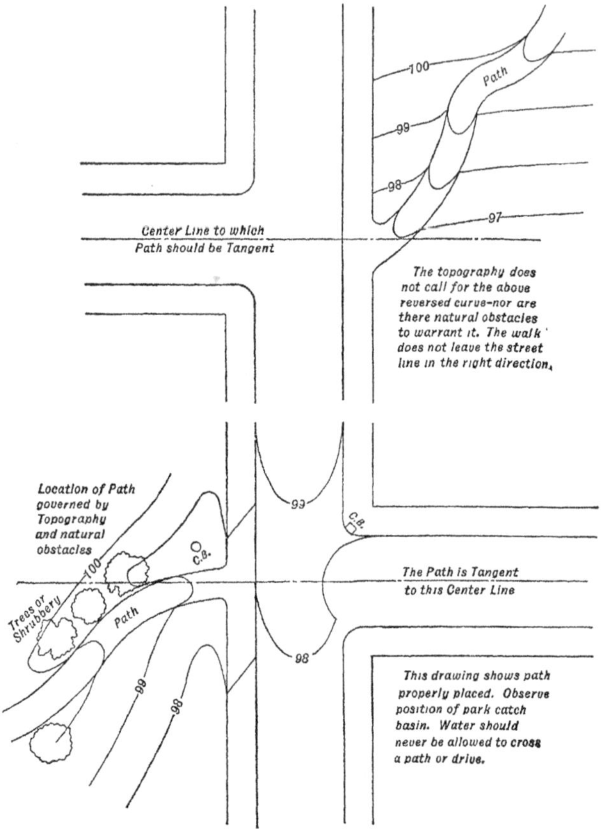
100
Path
99
98
97
Center Line to which
Path should be Tangent
The topography does not call for the above reversed curve–nor are there natural obstacles to warrant it. The walk does not leave the street line in the right direction.
Location of Path governed by Topography and natural obstacles
99
C.B.
C.B.
100
Trees or Shrubbery
Path
The Path is Tangent to this Center Line
99
98
98
This drawing shows path properly placed. Observe position of park catch basin. Water should never be allowed to cross a path or drive.

Fig. 23.

being of convenience in passing and when the walks are crowded. The depth of the gravel is 4 inches. The gravel should be of such a character that it will readily bind. It should be of suitable color and its surface should be continuous with that of the middle portion of the walk. Crushed stone will not answer for this purpose.

(4) *Entrances.* Where a park walk joins an exterior walk on the center line of a street, the street center line produced should coincide with a tangent to the center line of the walk at the point of entrance. If so designed the entrance will look correct to a person driving down the street toward it and will afford equal ease of access to the park in both directions. (See Fig. 23 on page 66.)

Corner entrances should be so laid out that they may be equally easy of access from both street cross-walks. This is illustrated in Figs. 24 and 25, on page 68, the first showing the correct design and the second the incorrect.

Cement Paths. Interior walks are crowned in the middle. The crown is of two kinds, straight and curved. The first consists of two planes intersecting on the center line of the path; the second of a continuous curved surface with high point in the center as before. The first method is to be preferred, as there is less likelihood of puddles after a shower. (See Fig. 15 on page 47.) Crowns are usually made too high. A crown of $\frac{3}{4}$ of an inch in a walk 8 feet wide is ample if it is actually secured. The importance of careful staking out cannot be over-emphasized. A little extra outlay for more grade stakes will more than pay for itself. It is in this connection that a criticism can be made in regard to the construction of our rural highways. In the effort to provide thorough drainage most extravagant crowns are adopted—crowns which frequently endanger traffic. By the employment of a small surveying corps

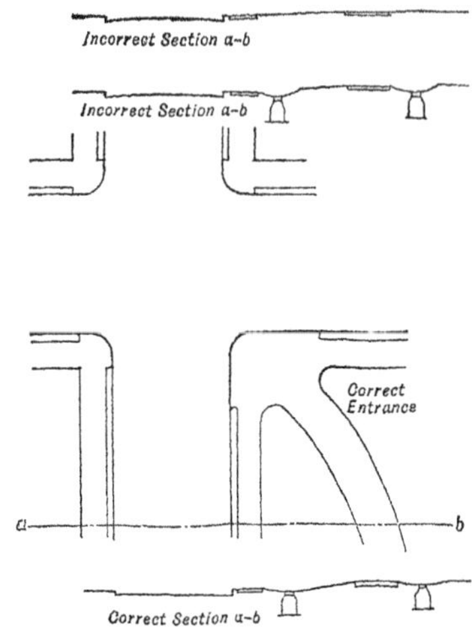
Incorrect Section a–b
Incorrect Section a–b
Correct
Entrance
a
b
Correct Section a–b

Fig. 24.

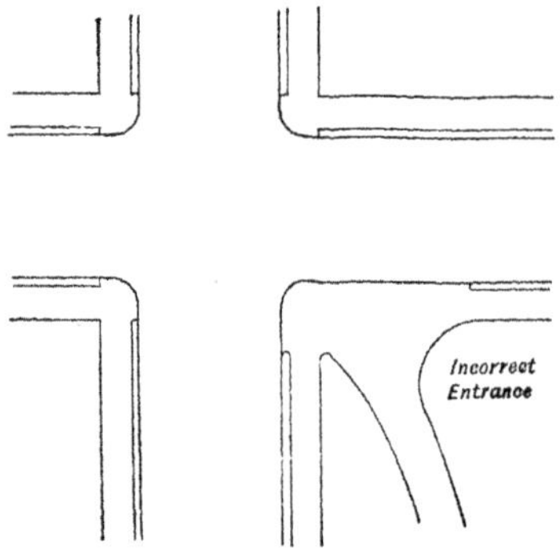
Incorrect
Entrance

Fig. 25.

in the construction of these roads, the crown could be greatly reduced and equally good drainage secured.

Exterior walks, that is, sidewalks along streets, have no crown, but are raised on the inside to provide drainage toward the gutter. A slope of 1¼ inches for an 8-foot sidewalk is sufficient.

The foundation of a cement walk consists of 8 inches of cinders. On this is placed 3 inches of concrete and on top of this 1 inch of rich mortar. The foundation is sometimes omitted in the South and may be occasionally omitted in the North if the walk rests on a bed of dry gravel. (See Fig. 19, page 55.)

The surface should be divided into blocks to provide for cracks. For an 8-foot walk, there should be a longitudinal marking down the middle and transverse markings about 4 feet apart. It is not wise to allow a smooth-troweled finish, as then the walk is apt to be slippery. A rough finish is better; but the roller finish is not to be recommended. Little advantage is gained by the undulations in the surface thus made.

ravel Paths. The crown should be a little higher than for cement walks, say 50 per cent. more.

The foundation should consist of 6 inches of cinders and the gravel should be 4 inches deep.

The selection of the gravel is a very important matter. Crushed stone is not to be recommended, but rather a hard rounded gravel with cementitious properties similar to the famous Roa Hook gravel found on the Hudson River. The gravel should of course be screened and should consist of round flat stones, not spheres. Gravel of this character is much easier to walk upon than gravel composed of spherical stones. Gravel is an excellent material for path construction. It is cheaper than cement, brick and asphalt and affords a very good footing; furthermore its appearance is less artificial. The Roa Hook gravel in particular is of a color which harmonizes well with the natural features of the park.

Asphalt Paths. This form of pavement is but seldom used. The crown should be as specified for the cement pavement. The foundation is of concrete 3 inches thick. On this is placed a binder course $\frac{3}{4}$ of an inch in thickness and on this the surface coat also $\frac{3}{4}$ of an inch thick. The amount of paraffin-petroleum residuum used as a flux in the surface coat should be a minimum in order that the pavement may be as hard as possible.

Brick Paths. These also are rarely used. The brick may be laid on a concrete foundation with a sand cushion between or may be laid on a bed of cinders leveled off with a layer of sand. The herring-bone bond is preferred.

DRIVES

The crown of park drives should be very low, so that the full width of the drive may be effective. For a drive 40 feet wide the crown should be about 9 inches. Instead of having a curved profile in cross-section, intersecting planes are recommended as for paths. This permits of driving on the extreme sides of the road.

Wherever possible, surface water should be carried in grass gutters and not in gutters constructed in the drive itself.

The effective width of park drives is frequently increased by the construction of wings on both sides. These are made thinner than the middle portion of the drive.

Park drives are usually constructed on curves rather than on straight lines, although parkway and boulevard drives are sometimes straight. (See Fig. 13, page 41.)

The Macadam Pavement. The telford pavement has been but seldom used, since the driving in parks has been light and the heavier telford foundation has been found unnecessary. It may be employed in the future owing to the increased weight of vehicles. The park macadam pavement has been constructed much lighter than the municipal pavement. Its thickness is 6 or 7 inches.

Heavier pavements are now coming into use. The ordinary water-bound macadam pavement often answers very well. Its cost is low and it affords a good foothold for horses. Where automobile traffic is heavy it will be necessary to treat the surface by the usual methods with crude petroleum, asphalt road oil or other preparation described at length in modern treatises on pavements. Bituminous concretes are giving excellent results for park pavements. Calcium chloride has met with some success in the East as a dust palliative.

Heavier pavements will be used in the future. The macadam pavement as heretofore constructed serves very well for carriage driving, but must now be treated with crude petroleum, tarvia, asphalt road oil or other similar preparation if heavy traffic is anticipated.

The Brick Pavement. This pavement possesses marked advantages in durability, but has the drawback of being noisy under horses' hoofs and steel tires. It has been used at Indianapolis for the construction of an automobile speedway. It is a more expensive form of pavement than the macadam, but is permanent. The maintenance charges for this style of pavement are very low.

Curb. Brick drives are always lined with curb and macadam pavements are occasionally. For brick pavements the curb may be either raised or flush with the surface. In the case of macadam pavements it is always raised. If the curb is raised the water must be carried in gutters. Curb used in park work is most generally of the curved variety and is very handsome when carefully laid, but extremely ugly when not. It is almost needless to say that curved curb is very expensive. Blue stone is a good material to use. Curved curb can be made of concrete with gutter of the same material. There is no good reason why curved concrete curb should not present an excellent appearance. This means that the engineer in charge of the work shall be a man of good judgment in laying out curves. Stakes should be set not farther than 50 feet apart. The adjustment of the scantling which are used as forms for casting the curb can be done by the eye. Considerable readjustment of the scantling back and forth may be necessary to secure the best results.

LIGHTING

Various systems are used and all have their merits. They are, 1. Electric Light; 2. Gas; 3. Acetylene; and 4. Gasoline. The lighting system should be as inconspicuous as possible and all standards and lights should be in keeping with the surroundings.

1. *Electric Light.* This method of lighting is the most common. All wires must be out of view. They are carried underground in lead conduits. The lights may be of the enclosed arc or the incandescent type—the former are usually preferred. Various kinds of ornamental iron posts are used, the wires being carried up inside the post.

Either direct or alternating current may be employed, whichever is available. If the park is at some distance from the supply, the alternating current will be found to be cheaper.

It is usual to provide distribution stations where the

current enters the park, so that in case of accident, the entire park will not be thrown into darkness.

Arc and incandescent lights may be connected either in multiple or series. It is possible to use both kinds of lights on the same circuit.

The multiple system of wiring is generally preferred, especially when the number of lights is large, inasmuch as it is safer than the series system. The voltage commonly employed for multiple wiring is 110, though it may be 220 and even 550, as in the case of a park illuminated by a trolley current. Willow Grove Park near Philadelphia is an example. To determine the voltage required for an arc light circuit in series, the rule is to multiply the voltage of each light, which is about 80, by the number of lights. Series wiring is frequently employed in small city squares.

2. *Gas.* The plain jet does not give sufficient illumination. Lights with mantles give a bright and satisfactory light. Gas can be used to advantage in districts where the price is low, as in the natural-gas belt. Gas mains should never be laid near trees and shrubbery. Leakage of gas exerts a very harmful effect on vegetable life.

3. *Acetylene.* This gives a brilliant and satisfactory interior light, but is seldom used in park lighting by reason of the expense. The acetylene gas is piped in wrought-iron pipes.

4. *Gasoline.* Gasoline lighting is cheap and satisfactory in several respects. The great drawback lies in time consumed in igniting the lamps. Each lamp is provided with a mantle, and a torch must be used to vaporize the gasoline.

CHAPTER IV

LABOR AND CONTRACTS

Day Labor versus Contract Work

THIS is a broad question. A few general remarks, however, can be made in this connection which may be of service.

Park work involving the use of elaborate machinery can be done cheaper by contract. This applies to such operations as steam-shovel work, pile driving, bridge construction and the erection of large buildings.

Grading work and masonry construction can generally be done cheaper by contract, as can sewer construction and the laying of water pipe.

The smoothing of the topsoil and its preparation for planting should be done by day labor if a well-organized superintendent's force can be gotten together. All gardening operations can be done much more satisfactorily by day labor.

Occasionally, however, some of the heavier engineering operations may be executed by day labor under good organization. Grading, path and drive construction, and even light masonry work have been done in this way.

It is a well-known fact among contractors that they can get much more work out of men than can municipal and park departments. The reason is obvious. The contracting firm is better organized—usually with one man at the head who gains or loses financially on each contract. A laborer who is inefficient is immediately discharged.

By the other method, the work could be carried on satisfactorily if the foreman were free to discharge an inefficient workman. But this often is not the case, and the foremen themselves are frequently employed on the recommendation of some official or prominent citizen.

Nevertheless the day labor method is coming more and more into vogue and appears to be giving good results when politics can be eliminated, as is often the case. The claim is made that the work can be done cheaper by day labor than by contract, since by the former method there is no profit to be included. Furthermore the park superintendent, by offering permanent employment in one locality, is able to attract to his force a class of steady labor to whom the contractor's inducements do not so strongly appeal.

TWO KINDS OF CONTRACTS

Park contracts are let either by the lump-sum or by the unit system.

1. *The Lump-sum System.* By this method but one price is stated in the bid; namely, the cost for the entire work. This system is inelastic, but is used when the work in exact nature and amount is fully known before the contract is let. If the work is not fully known beforehand there will surely be claims for extras on the part of the contractor.

2. *The Unit System.* This is the method usually employed in engineering. The work to be done is divided into different parts, such as earth excavation, sewers, cast iron, macadam pavements, etc. Approximate figures representing the amount of work to be done under each item are stated beforehand and each bidder submits his bid on the price per cubic yard of earth excavation, the price per lineal foot of sewers, the price per pound of cast iron, the

price per square yard of macadam pavement laid complete, etc., basing his total estimate on the above approximate estimate as given by the engineer. This system is elastic. The amounts of work under each heading can be increased or decreased, although not indefinitely. The increase or decrease permitted is usually not more than 20 per cent. If a contractor has bid low on earth excavation and high on sewers, he naturally would object to the amount of earth excavation being greatly increased or the lineal feet of sewers greatly reduced. Under the unit system it is not necessary to prepare complete plans of the work before it is begun.

METHOD OF PROCEDURE

In contemplating the execution of a piece of engineering work, three questions must be decided by the board or commission having power. These are, first, What will be the cost? second, Is it advisable to do it? and third, Shall it be done by day labor or by contract?

In determining the first question the engineer is consulted and asked for a report on the cost. This report may be made immediately in some cases, but in others considerable time may be required for its preparation.

After knowing the cost, the advisability of doing the work is discussed and a decision arrived at.

The question as to whether the work shall be done by day labor or by contract is usually referred to the engineer.

THE ADVERTISEMENT

An advertisement of the work is inserted in the local papers and sometimes in one or two of the leading technical journals. Unless the contract is to be let by the lump sum, the amounts of all the items are stated. The advertisement also states the time and place for receiving bids and the amount of the proposal bond.

THE CONTRACTOR'S EXAMINATION OF THE PLANS

If the contract is a lump-sum contract the plans must be complete and perfect in every detail. If a unit-system contract, it is not necessary that the plans be entirely complete.

These plans are usually blue-printed and prints are sometimes furnished to prospective bidders either free or for a deposit which is returned when the prints are returned. Sometimes no prints are furnished for distribution and prospective bidders are required to examine the plans in the office of the board or commission. Then again blue-prints are made on a reduced scale and are furnished free in booklet form to all responsible applicants. This method is up to date, convenient and cheap. Should any parts of the prints be too small to be perfectly intelligible, the full-size drawings can be examined at the office. The scales on all drawings which are to be reduced by photography must be graphical scales so that they may be reduced in the same ratio as any line on the drawing.

It is advisable to furnish prospective bidders with full information in regard to the work. This will reduce the amounts of the bids. It is the universal practice of bidders to raise their bids where the element of uncertainty enters in. For this reason in sewer work it is advisable in many cases to make complete borings and prepare profiles, so that bidders may know just what they are likely to encounter. There may be a distinct understanding in the contract, however, that the party of the first part will not hold itself responsible for errors in said profiles nor become liable for errors. Similarly, in other work, as much information should be furnished as possible.

Prospective bidders should be advised to make a personal examination of the site of the work and they should be notified that they themselves are held responsible for the accuracy of the information furnished by the party of the first part. It should be said, however, in this connection, that courts of law may not support the party of the first part in its effort to relieve itself of responsibility, especially where gross blunders have been made.

Each prospective bidder is given a proposal blank and a set of specifications. The proposal blank is a form in which each bidder inserts his prices and fills out an agreement, stating that if he is awarded the contract he will accept it. The proposal also contains the bond to secure the agreement.

The specifications, so called, are the specifications proper. for all of the different kinds of work to be done under the contract; and also include a blank form similar to that in the proposal to be filled in by the successful bidder, together with a contract bond to secure the agreement. As soon as the contract is duly executed, the proposal bonds all become void.

The various articles in the specifications stating how the work is to be done should of course be suited to the particular undertaking in question. The "paste-pot and scissors method" used by some engineers is not above criticism. Then, too, there are engineers who have a way of writing what may seem to be unnecessarily strict specifications to protect themselves against contractors on the lookout for loopholes in the agreement. Contractors soon learn the personality of an engineer and in preparing their bids take it into consideration.

THE CONTRACTOR'S CALCULATIONS

These are made after all the necessary information has been gathered in. Many contractors employ an engineer to make their computations and to fill in and submit the bid. Careful bookkeeping of costs and good system on the part of a contractor and his engineer yield good financial returns. Many contractors who do not take the trouble to accumulate cost data think that they know what the expense to them for various kinds of work will be, whereas they do not. This is especially so in such matters as earth excavation. Lack of system in preparing cost data on this one item alone has led to failures and ruin. The cost of removing a cubic yard of earth is a function of many component items, such as superintendence, plowing, picking, shoveling, carting, spreading, blacksmithing, etc. The cost of each of these items for various classes of material should be accurately known.

THE SURETIES

There are four ways of securing a proposal or contract, and in addition there is another form of insurance called Contractor's Liability Insurance, all of which will be discussed.

1. *By the Bidder's Reputation.* This may be said to be no surety at all in the strict sense. In small places and for unimportant work sometimes no bond is required. Should a bidder refuse to accept a contract awarded him, his reputation in that community is practically ruined and his hopes of securing more work of that or some other kind are practically destroyed.

2. *By Certified Check.* For work of no great magnitude, this method is often followed. Each bidder submits with

his bid a certified check in the required amount. These certified checks are returned to all except the successful bidder and his check is returned when the contract is executed, the contract being secured by a contract bond, described under headings 3 and 4.

3. *The Personal Bond.* Proposals and contracts may both be bonded in this way, though this kind of bond is not now in frequent use. The proposal may be thus bonded and the contract bonded by a surety company (see 4). The amount of bond is stated in the advertisement and may be equal to or less than the total cost of work. Thus, if the amount of the work is $100,000, each bondsman, there often being two, makes affidavit that he is worth $50,000 above all his debts and liabilities of every nature.

Personal bonds for the contract may be executed by the same men who acted as bondsmen in the proposal or they may be executed by others. When a bondsman signs a proposal bond he is liable also for the contract should it be awarded to the bidder for whom he acted as bondsman. This obligation, however, is practically never taken advantage of, nor abused.

4. *The Surety Company Bond.* The obtaining of a surety company bond is purely a matter of business and thus is much more satisfactory than the individual bond, where obligation is always incurred and where reciprocal favors are usually asked if no charge is made for the bond. A surety company before acting as bondsman for a contractor makes searching investigation of his financial standing. If this is satisfactory the bond will be executed, for which a percentage payment is required. Surety companies do not necessarily make their investigation before the execution of every bond, but use their discretion in the matter. When an applicant has once thoroughly satisfied

a surety company of his financial standing and ability as a contractor he has little or no trouble in getting bonds whenever he wants them unless the conditions are peculiar as to the amount or other attendant circumstances. Some contractors never use individual bonds and others only use them on proposals.

5. *Contractor's Liability Insurance and Compensation.* Insurance protects a contractor from suits for damages arising from personal injury or death to a workman or to an outsider. It thus protects a contractor from financial failure and thereby protects the bondsman as well. A surety company may require a contractor to obtain liability insurance before it will act as bondsman. There are companies who make this type of insurance a specialty. The rates charged are proportional to the risk run and the magnitude of the work. They are higher for instance in heavy sewer work than they are in ordinary excavation.

Several of our eastern states now have "Workmen's Compensation Laws," which provide for a compensation to be paid to an injured workman or to his family in case of death. Such a law has recently gone into effect (Jan. 1, 1916) in the state of Pennsylvania. Whatever payment a contractor may be required to make to purchase insurance against payments of compensation will have to be included as one of the overhead charges and will increase the cost of the work accordingly.

MAKING OUT THE BID

This is done by the contractor or his engineer. All blank spaces in the proposal are filled in, the bonds duly executed and the bid enclosed in a sealed envelope.

SUBMITTING THE PROPOSALS

These are submitted to the board or commission, preferably publicly, at the time and place assigned. If the bids are thus submitted, the clerk of the board reads them aloud. If the bids are by the unit system, some minutes (or perhaps longer) will be required for the engineer to determine the order of the bidders. To do this he multiplies the price bid for each item by the number of units of work as determined by the advertisement and then sums up the products.

REJECTION OF BIDS

The party of the first part reserves the right to reject any or all bids. The contract is awarded to the low bidder providing he is able to "qualify," unless his bid is seriously in excess of the amount of the preliminary estimate of cost. If, however, the low bidder is not acceptable either financially, personally, or in the matter of experience, the next low bidder may be awarded the contract providing his bid is satisfactory in amount. The rejection of the low bid is apt to give rise to adverse criticism.

AWARDING THE CONTRACT

The contract is often awarded on the same day on which the bids are received. Sometimes, however, in order that various matters may be considered, such as the responsibility of the low bidder and the cost of the work, the awarding of the contract is postponed.

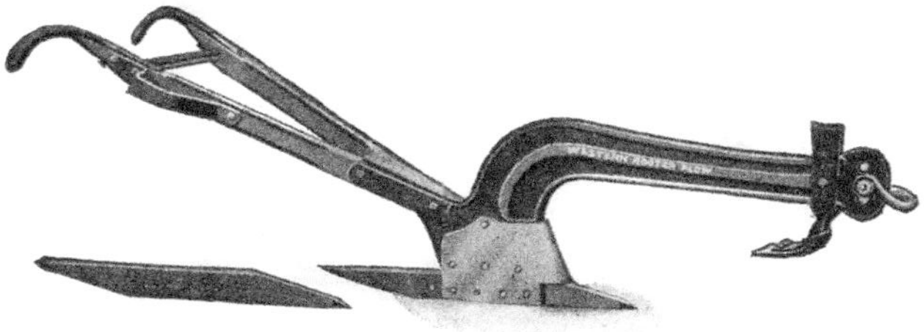

Courtesy Austin-Western Road Machinery Co.

Fig. 26.—Rooter Plow.

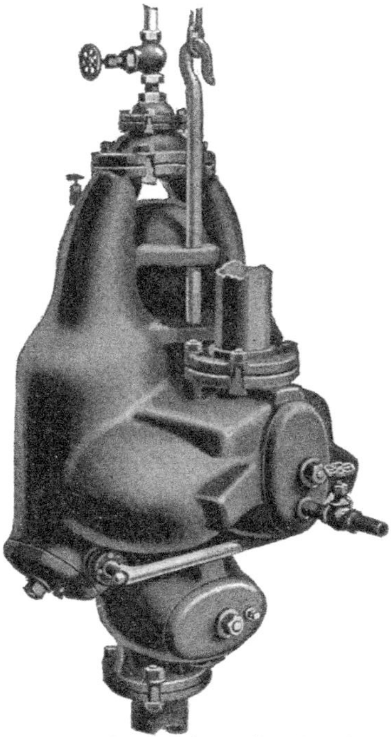

Courtesy Pulsometer Steam Pump Co.

Fig. 28.—Pulsometer.

Courtesy Austin-Western Road Machinery Co.

Fig. 27.—Wheel Scraper.

EXECUTION OF THE CONTRACT

The bidder to whom the contract is awarded is require to execute the agreement within a specified time. To d this he must affix his signature to the contract form in whic all of the prices have been entered by the clerk of the board The agreement is secured by individual bonds or a suret company bond, which constitutes a part of the contract As mentioned before, a surety company usually require a contractor to take out liability insurance before it exe cutes the contract bond. After the contract is signed an the bond duly executed all of the proposals together wit the accompanying bonds become at once void.

STARTING THE WORK

Active operations must be begun within a specifie number of days, which is frequently ten.

THE PAYMENTS

Monthly Estimates. At the end of each month it i customary in many contracts to make a payment, but no in the whole amount of the work done. Sometimes 1 per cent is retained and sometimes 25. It is not nec essary for the engineer to make an accurate determinatio of the quantities, although, if to be on the safe side h underestimates the work, the contractor may be dissat isfied, especially if his payroll and other obligations ar heavy.

The Final Estimate. When the work is satisfactoril completed a final estimate is prepared which is as exact a it is possible for the engineer to make. The final paymen is equal to the total value of the work less the sum of th previous payments.

EXTRAS

These, as are well known, give rise to dissatisfaction and dispute. A contract should be so framed that the approved bill for extras will be as small as possible. Considerable care and foresight are required on the part of the engineer in order to provide in the contract for contingencies which would otherwise be overlooked. The adjudication of the bill for extras is difficult and disagreeable.

PENALTY CLAUSES

If the work does not progress at the proper rate, it is the duty of the engineer to give notice to that effect.

The "penalty clause" in most contracts imposes a certain charge per day for non-completion of the work on contract time. Provision is made in the specifications that the penalty be deducted from the moneys due the contractor on the final estimate. These clauses are apt to be very faulty from the legal point of view, notwithstanding the fact that, although written by the engineer, they are passed on by the counsel (see page 14). Contractors, as a rule, who know the law in such matters better than engineers, pay but little attention to these "penalty clauses," knowing that they are nearly void. In order to overcome the legal difficulty involved in a penalty, the clause sometimes states that the charge made for non-completion of the work is not to be construed as a penalty but as liquidated damages. This clause will also not be upheld in court unless it can be shown that the damages claimed are those actually sustained. Courts of law in general are opposed to hard-and-fast agreements made in advance to cover damages which may be sustained for non-completion of the work on time. If, however, these

Fig. 29.

agreements are made so elastic as to adapt themselv
to the individual case they will be upheld. This end m
be attained by itemizing the damages as so much per d
for engineering services, inspectors, clerks, office expens
etc. An agreement made with the full understanding
both parties to the contract, stating that, in view of t
difficulty of estimating in advance the exact amount
damages involved for non-completion of work on contra
time, a certain sum per day is agreed upon as the act
liquidated damages as nearly as they can be estimat
will sometimes be upheld.

CHAPTER V

CONSTRUCTION

BEGINNING OF OPERATIONS

THE time for starting the work is stated in the specifications. The time allowed is often "not later than ten days after the awarding of the contract." The necessary machinery must be brought on the ground at once, and stables, shanties and offices provided. If there are any suitable buildings on the land, they may, with the engineer's permission, be used as stables and offices. In order that these buildings may be available until the completion of the work, it is desirable that their location be such that but little work shall remain to be done after their removal. In parks remote from town, the contractor is permitted to erect shanties for the men. Their needs are supplied by the company's store which, though necessary, needs regulation in its methods in order that the interests of the men may be protected.

UNDERDRAINAGE

Before any other operations can be accomplished on swampy lands, the ground must be drained in order to lower the water-table. The laying of tile drains should be begun at the outlets, although in order to hasten the work operations may be begun at several parts of the system. Where lines of tile are laid without an outlet, water will always collect at the starting-points and this

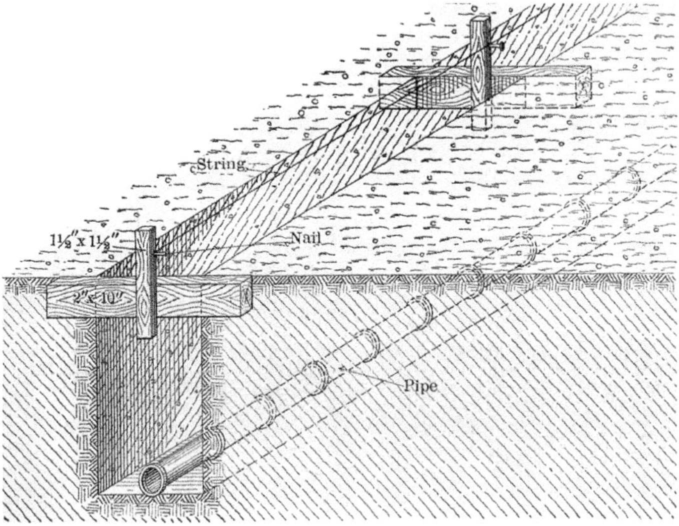

Fig. 30.—Line and Grade for Sewers.

Fig. 31.

must be pumped out before the work can be joined. Th ditching is done by pick and shovel. The shovels may b either short-handled or long-handled. The former ar generally better, as the men are more familiar with thei use. Grade is given by stakes driven on the side of th trench on which the cut from the top of the stake is marke in feet and inches. These stakes are driven every 50 fee except for main ditches, when they should be 25 feet apart The minimum grade for mains is 1 inch per 100 feet and fo laterals 3 inches. The trenches are hollowed out on th bottoms for the collars so that the weight on the tile ma not be borne at the ends. Tile drains should be lai below frost and out of reach of rootlets, particularly o willows. Professor Mapes, who is quoted by Colone Waring, specifies that drains must be 3 feet deep when 2 feet apart, 4 feet deep when 40 feet apart, and 5 feet dee when 80 feet apart.

In backfilling no stones should be placed closer than inches to the tile and the subsoil should be thrown i first. Some engineers require that in excavating the trenc the topsoil be placed on one side and the subsoil on th other. The backfilling is tamped in 9-inch layers to secur the best results.

SEWERS

Line and grade are best given by placing 2-inch b 10-inch planks transversely in the ground every 50 feet These planks do not have to be set at any exact elevation but they should be firmly embedded and should be of suffi cient length so that when the trench is excavated the will remain immovable. The center line can be locate by driving a tack on top of the plank. After the excavatio work is partially finished a 1-inch by 2-inch strip is naile vertically on the plank with one edge exactly on the center line of the trench. This strip must be plumb. Then by the use of a wye-level, a mark is made on the side of strip and a nail driven a whole number of feet above the invert elevation. This is also done at the next 50 feet and a string stretched between the nails. The string may thus be stretched for some considerable distance at the given number of feet above the invert grade line. Errors in level work are easily detected. If by reason of an abrupt change in cut the line strikes the board, or too high or too low above or below it, the string may be lowered or raised a foot or two. The usual methods of pipe-laying and inspection should be followed and will not be discussed in this connection. (See Fig. 30, page 91.)

Wet trenches can be kept free from water by use of a hand-pump or the pulsometer (see Fig. 28, page 83). In park work a hoisting engine is generally available and connection may be made between the pulsometer and the boiler of the hoisting engine, a low fire being kept up over night by the night watchman. In the morning the trench will be in a suitable condition for work.

The backfilling should be tamped in layers. If the main line trench is filled in before the laterals are laid, a plank is placed in a vertical position against each wye connection and sufficiently long to project above the ground.

GRADING

Staking Out the Work. Grade stakes are set at the intersections of the cross-section lines and at every change of slope. The stakes are marked on the side with the proper cut or fill measured from the top of stake or they are so driven that their tops are to grade, in which case the tops

Fig. 32.

are marked with red water-proof crayon. Stakes set t grade are better than cut or fill stakes and should be use wherever possible even though a slight hole may be required Laths may be driven alongside to indicate their positions Cut and fill stakes should be marked in feet and inches an not in feet and tenths. For heavy fills long poles ar sometimes used with their tops at the finished grade, thu doing away with the labor of setting more stakes after th first have been covered up. Where the cut is considerable the first cut stakes set will not remain during the excavatio operations unless left on little hillocks, and to do this i expensive. Therefore after the first cut stakes have bee plowed up or otherwise removed, other stakes in the sam position will be necessary.

Topsoil Stripping. The first thing to be done is t strip the entire tract of its topsoil, which may be from inches to a foot in depth. The soil is piled in large mound with spiral drives up the sides in positions where there i to be little cut or fill. It would appear to the layma cheaper to excavate a certain part of the tract to th proper subgrade and then to cover it with topsoil strippe from an adjacent tract. This, however, may not be th case, and to systematize the work and avoid confusion, th entire site is stripped at the beginning of operations. I large operations the topsoil is piled; in small ones the top soil may be spread direct, thus saving the cost of doubl handling of the material. (Figs. 7, page 29, and 10, pag 37.)

Excavation by Pick and Shovel. The soil is broke up by picks and shoveled into dump carts, skid wagons o dump wagons by the use of round-pointed, short-handle shovels. From the engineer's point of view this method i the most satisfactory, because the subgrade can be dresse more accurately and because the grade stakes can all be preserved until the finished grade is reached. Carts are used for short hauls and wagons for long hauls.

Excavation by Plow and Shovel. By this method the soil is broken up by the pick- or rooter-plow, whose point is much heavier and longer than that on an ordinary agricultural plow. (See Fig. 26, page 83.) Carts or wagons are used as in the preceding case. It is of course much cheaper to plow the soil than to loosen it by hand providing that the work is of such a character that the plow is available.

Excavation by Grading Machine. The Austin Excavating Grader is a machine well adapted to topsoil stripping over broad areas. This machine is driven by 12 horses, 8 in front and 4 behind, or by a traction engine. It consists of a heavy plow which throws up the soil on a revolving belt which carries it up and then drops it in a dump wagon driven alongside. At least two drivers are required and sometimes four at the beginning of operations in order to properly manage the horses. One man is needed to operate the plow. (See Fig. 11, page 37.)

By Steam Shovel. Where the cuts are very heavy and the work large in amount, it will pay the contractor to use a steam shovel. He will also need a few light locomotives and an outfit of dump cars, and probably one or more centrifugal pumps to keep the pit where the shovel is located free from water. The filling is done from trestles. The settlement will thus be much more than by the other methods where the fill is consolidated by wheels and horses' hoofs. It is not advisable to attempt road construction on filling thus made until thorough settlement has taken place, nor to lay sewer and water pipes, unless tamping has been resorted to.

Fig. 31, page 91, furnished by the Essex County Park Commission, shows a shovel making the first cut in the Middle Division of Branch Brook Park. This work was necessary in order to create an artificial lake.

In park work much water is encountered in the excavation for lakes, which must be removed through sewers or else by low-lift centrifugal pumps. Fig. 32, page 95, shows the same work at Branch Brook Park in a more advanced condition. This work was particularly difficult by reason of large volumes of surface and ground water.

By Dredge. Where large bodies of water must be deepened some form of dredge will be necessary. The accompanying illustration shows a clam-shell dredge removing earth and vegetable growths at Weequahic Reservation, one of the Essex County Parks. This lake is adjacent to the Newark Meadows (salt marshes) and originally was quite shallow. It was increased to a depth of 8 feet in order to prevent the growth of water plants, the excavated material being largely used for surface dressing. (Fig. 33, page 101.)

Fig. 34, page 105, shows a hydraulic dredge at work at Lake Nokomis, Minneapolis. This dredge is mounted on a timber hull, 80 feet long. It is electrically operated by an alternating current of 2300 volts. The main pump has a 15-inch suction and discharge, the pump revolving at either 250 or 305 R.P.M. as desired. The two winding shafts each run at 22 R.P.M. There are five drums used as follows: one each for the right and left spud, one each for the right and left swings, and one for raising and lowering the suction ladder. The ladder and cutter dig to a depth of 28 feet. For a greater depth a plain suction sleeve is used. The dredge is provided with 1600 feet of pontoons with pipe and also 4000 feet of shore pipe. The illustration referred to was furnished by Mr. J. A. Ridgway, Secretary of the Board of Park Commissioners, and the description is from the 32d annual report.

By Drag Scraper. For very short hauls, not above 100 feet, the drag scraper can be used to good advantage. The soil must first be broken by plow or by pick. This work is very hard on the horses, as they are not allowed the regular breathing spell which obtains in all other work with the exception of that done by the excavating grader and wheel scraper.

By Wheel Scraper. For hauls from 100 to 300 feet the wheel scraper may be used providing that the ground is level and the cut is fairly uniform. No other class of work is as hard on the horses as this, and drivers should be instructed to stop them occasionally. A bucking team must sometimes be hitched on the pole in front of the other team in order to load the scraper. (See Fig. 27, page 83.)

At Westside Park, Newark, N. J., peculiar methods were adopted for the excavation of an artificial lake by reason of the unusual character of the material removed. Below a surface crust of topsoil about a foot thick was encountered a deposit of muck consisting entirely of vegetable matter, absolutely free from sand and clay. The excavation reached a depth of 8 or 9 feet, though in some places the muck extended downward for 35 feet. The excavated material was used as topsoil and supported phenomenal growths after a year's decomposition. To remove it three methods were adopted—first, by means of derricks; second, by means of tram cars; and third, by the assistance of a traveling cableway. In Figure 35, page 109, are shown three derricks with the black muck in adjacent piles. This of course had afterwards to be distributed over the finished subgrade. In the background is

Fig. 33.

seen a pile-driver at work on a timber bulkhead to suppo the farther shore. This bulkhead was constructed accordance with the general description on page 54.] the foreground is seen part of a shore wall along the near shore. The exceedingly unstable character of the sc can be seen from a close examination of the houses in tl background, several of which are out of plumb. In Fi 36, page 113, is observed the second method of excavatio as mentioned in this paragraph. The pit was kept fr from water by means of a pulsometer as shown.

After the lake excavation was well along, good uplar subsoil was dumped over the timber bulkhead as describe in general on page 54. In sinking to a condition of equilil rium a considerable quantity of muck was thereby di placed. This was all removed by means of a cablewa and buckets, since the bottom of the lake had becon exceedingly soft and was incapable of supporting tracl or run-ways.

The Essex County park lands include a great mar acres of originally poor and wet soils. These have bee developed into lakes and water courses, thereby great enhancing the beauty of the parks. On the other han by means of judicious filling, marshes and areas of shallo submergence can be transformed into valuable pai property. An example is seen in Figs. 3 and 4 on pag 15 and 19, loaned by Commissioner Cabot Ward of tl Department of Parks of the City of New York. Tl filling, as stated by Acting Commissioner Valentine, w largely deposited under permits issued to excavators ar contractors. At the upper end of Riverside Park Extensic a good deal of the fill was obtained from excavations f the Catskill Aqueduct.

Mr. G. A. Parker, Superintendent of Parks of Hartfor

Conn., has made a careful study of the time element in excavation work by shovel, which he has presented in a paper in the Proceedings of the Connecticut Civil Engineers and Surveyors' Association. His theory as far as the laborers are concerned is divided into three parts:

First. That a shoveler can do maximum work when he shovels 5 shovelfuls in 50 seconds. It takes each man about 5 seconds for one shovelful. He will therefore work 25 seconds and rest 25 seconds, and so on throughout the day. The rest periods will be his absolutely without any interference on the part of the man in charge of the work. Experiments were made on 4 shovelfuls in 40 seconds, 6 shovelfuls in 60 seconds, 10 shovelfuls in 100 seconds, also, the result being that 5 shovelfuls in 50 seconds gave the greatest amount of work. By this method, after deducting 5 per cent. for waste time, Mr. Parker claims that an average laborer can handle $22\frac{4}{5}$ cubic yards in a 10-hour day.

Second. That 4 men, 1 working alone and loading 10 yards in a day, 1 working in a gang of 5 and loading 12 yards, 1 working in a gang of 10 and loading 15 yards, and 1 working in a gang of 30 and loading 22 yards, are equally tired at the end of the day. It is Mr. Parker's opinion that laborers on this class of excavation should work in gangs of 30 men each, subdivided into crews of 15, one crew for every cart.

Third. Each man is required to count his shovelfuls. In each 50 seconds he must load 5 shovelfuls, no less, no more. The claim is made that the counting has a beneficial and stimulating effect.

FIG. 34.

PILE DRIVING

Wooden Piles are generally of spruce, pine or oak. Hemlock does not possess sufficient elasticity. If they are to be used as columnar piles it is necessary that they be absolutely straight; if as bearing piles they should be fairly straight; and in either case of sound timber. The minimum diameter of the butts is 12 inches and of the tips 8 inches. Piles should be sharpened to a point before driving and sometimes it is necessary to protect the point by an iron shoe. The butts are cut off square and adzed so that a wrought-iron band may be slipped on to protect the pile in driving. Piles should be driven until the movement is not more than one-half an inch under a 1500-pound ram falling 15 feet on the last blow.

Piles are used in park engineering in constructing foundations for walls, abutments, spillways, etc. When used in foundations it is better to cap them with concrete than to attempt to construct a timber grillage.

The land machine traveling on rollers is the common type. Fig. 29, page 87, shows a land machine driving vertical piles for the construction of a timber and earth bulkhead. (See Figs. 35 and 36 on pages 109 and 113.) This machine is held upright by guy ropes attached to "dead men" or to posts set in the ground. These posts are sometimes twisted into the ground where the soil is not very hard. This method consists in attaching a long cross-piece to the post by means of a chain. One or two men at each end of the cross-piece by walking in a circle and pressing downward thus cause the post to sink into the ground. The ease, simplicity and effectiveness of this method is astonishing to those who have never witnessed it before.

Concrete Piles are of two general types—those that are

moulded before driving and those that are moulded afterward. Those of the first class must be reinforced. The Chenoweth and the Cummings and the Hennebique are pre-moulded piles, and there are also many unpatented types of the same class. Those of the second class are of various forms, as the Simplex, the Raymond and the Pedestal.

Concrete piles are much to be preferred when acting as columns, resting on a hard substratum. They are more durable than wooden piles in ground which will subsequently be dried out through the absorptive action of adjacent brick sewers and are thus especially suited for use in certain park lands. They are used to support arch abutments and masonry retaining walls.

MASONRY WALLS AND STEPS

Retaining Walls. One of the most important of the practical matters to be attended to is the excavation for the foundation. It is not necessary that the foundation should extend down to the bottom of frost. The heaving effect of frost can be entirely eliminated by digging down $2\frac{1}{2}$ feet below the finished grade.

The foundation pit should be excavated true with ample room on the front side for the construction of a good square toe. Skimping on the front side of the foundation has resulted in the failure of many retaining walls.

Walls for shore protection should be founded on good soil. The use of spud piles is not in general to be recommended. It is better to excavate to a hard foundation. The reason for this lies in the fact that excavation in front of a shore wall or sea wall permits fine sands, "liver" and quick-sands to flow away from under the foundation, thus endangering its stability.

Fig. 35.

The neat lines for retaining walls are obtained b setting " A-frames " in position by the help of a trans so that the neat line on top of the footing and the nea line under the coping can be located by stretching string from frame to frame or else from a frame to the finishe wall.

In order to secure a perfectly straight coping, especiall in long walls along city streets, it is suggested that a tra sit line be established on top of the wall before the copir is placed, say 2 inches back from the neat line. This lir can be marked by crow-feet every 25 feet.

Retaining walls at the foot of a slope are separated fro it by a swale gutter. The coping is so designed as to thro the water falling on it backward into the gutter, where it collected by catch-basins or inlets.

The matter of pointing is very important. The join should be thoroughly raked to a depth of at least 1½ inch before the pointing mortar is applied. Depressed join pressed in with a pointer's tool are much superior in dur bility to raised joints. Only the best Portland cemei should be used for this purpose.

WATER PIPES

The practice of laying park water pipes prior to tl completion of the grading work has sometimes been r sorted to. It has been claimed that this is advantageo in lands to be filled and no doubt it is from the contracto standpoint, as considerable digging is thereby avoide The practice, however, is open to serious objection, as some places the pipes are almost if not entirely uncover and the driving over them of wheel-scrapers, dump wagon etc., frequently seriously damages the pipe. Furthermo if the filling operations are not completed before winte the pipes are apt to freeze and burst, as they are seldom free from water in low places.

All pipe should be thoroughly tested before the backfilling takes place. This can be done by admitting city water into the system and examining the pipes for leaky joints, pin holes and split seams. If a higher pressure is desired the pipes can be filled with water and the supply shut off. Then by attaching a hand pump with pressure gauge to a small nipple tapped into the pipe the pressure can be raised as high as desired with the addition of but very little water. Sometimes where no water supply is available sections of the system must be filled by means of the hand pump. This is a very slow process for cast iron pipe but not very difficult with the smaller wrought iron sizes. All defective pipe should be removed and no plugging of holes should be permitted.

Water hammer is very likely to occur when new pipes are filled with water. To avoid this open up a few lawn hydrants, especially at the lower parts of the system.

PATHS

The excavation for path foundations should be properly formed and of full depth over the entire cross-section. The subgrade is then rolled and filled in with cinders to the proper depth. These cinders should be of the gas-works quality, sharp and clean. Ashes will not do. The cinders are then wetted by means of a sprinkling cart and thoroughly rolled. On the cinders is placed the pavement proper whether of gravel, cement, asphalt or brick.

Cement, brick and asphalt pavements are laid between 2-inch by 4-inch scantlings to hold them in position. These scantlings are held in place by means of stakes. As park walks are usually curved, the scantlings must be curved to

Fig. 36.

fit the plan. This curving is best done in the field by tl aid of the eye alone. By the exercise of due care in alignir and leveling the scantlings, high-grade work can be secure

The practice of using round field stone for gravel pat foundations is not to be commended, as the stone worl up to the surface. Gravel paths should be thorough wetted and rolled before being opened to the public.

Cement paths should not be laid in freezing weather u less the work is carefully protected over night by stra If the mortar freezes it will have to be removed. Tl surface coat, usually 1 inch in thickness, should follow tl 3-inch foundation layer of concrete before the latter h had time to dry out. A heavy rain falling on a fresh laid cement walk washes out the cement and leaves tl surface in a pitted and porous condition requiring remova

DRIVES

The Macadam Pavement. The first thing to be do in the construction work is to excavate to subgrade. Tl subgrade should be thoroughly rolled with a 10-ton roa roller. All soft spots which are revealed by the rollir process should be removed and the holes filled in with goo earth. The subgrade should be parallel to the finish surface. The foundation is then placed. Four inches 2½-inch broken stone are sufficient for a macadam and inches of telford foundation for the better class of road The macadam foundation should be carefully spread wit forks and sprinkled with coarse screenings and wetted an rolled. The telford foundation is prepared by placir stone at least 7 inches high in such a way that the edges a upward. These stones must all be placed by hand. A points projecting higher than 7 inches are napped off wit napping hammers. On top of the foundation is place

2 or 3 inches of $1\frac{1}{2}$-inch stone, which is wetted and rolled. The surfacing consists of an inch of screenings thoroughly wetted and rolled. Some engineers use a little earth binder between the courses and even on top of the finished pavement.

The best water-bound broken-stone road soon ravels and disintegrates if subjected to the destructive action of rapidly moving automobiles unless special preventive measures are resorted to. In order to protect roads already built and construct new roads in a more durable manner, the following methods are adopted:

1. *Sprinkling with Fresh Water.* As carried on in the past this has been very expensive and in municipal practice has often cost over $500 per mile per annum. Mr. S. Whinery, M. Am. Soc. C.E., strongly advocates the use of water under improved conditions which he admits are necessary to make this method a success. When the watering is properly done he is of the opinion that the wear from travel is no greater than when a light oil is used. By properly watering the pavement he states that raveling can be prevented. Furthermore there will be no dust if the pavement is constantly moist. Mr. Whinery advocates the use of power sprinklers instead of the old-fashioned water wagons—sprinklers which shall be driven by power and which shall deliver the water under a constant head regulated by a pump. The author is of the opinion that Mr. Whinery's method of sprinkling would be especially successful in parks where opportunities for carrying on the work under the most favorable conditions would be possible. Here in the park the sprinkling would be done by an employe under the supervision of a foreman or superintendent who would make it his business to see that the work was carried on efficiently.

2. *Sprinkling with Salt Water.* The results are more durable than fresh-water sprinkling, since the salts in the water are hygroscopic in their properties. Should the pavement become dry, the dust containing salt is thrown into the air and is extremely irritating to the throat and eyes.

3. *Sprinkling with Water and Calcium Chloride.* This salt is strongly hygroscopic and roads treated in this manner have proved quite satisfactory in England, where the climate is moist. The mixture to be used is 1 pound to 1 gallon of water, using $\frac{1}{3}$ of a gallon of the solution for each square yard. Ten applications per season are usually sufficient.

Calcium-chloride treatment has been adopted by the Metropolitan Park Commission of Boston and by the Department of Parks of Hartford, Conn.

Calcium chloride is also applied as a dry powder, the pavements being first swept by hand. The amount of chloride used varies from $\frac{3}{4}$ to $2\frac{1}{2}$ pounds per square yard.

4. *Sprinkling with an oil emulsion* consisting of water, oil and an alkali. Vegetable oils, crude petroleum and coal tar have been used in municipal practice. Potash, soda, ammonia and soap solutions are the mediums employed to render the oil miscible with the water. The mixture is sprinkled from an ordinary watering cart, though a power sprinkler as described on page 119 would give more satisfactory results. The surface may be left uncovered or else treated with a thin coat of sand.

5. *Treating the Surface with Light Oils and Light Tars.* The oils and tars are distributed preferably by pressure distributors. Those with hoods have been employed to protect pedestrians. The palliatives employed are vegetable oils, paraffin and asphaltic petroleums, tar oils, water-gas tars, coal-gas tars and various patented compounds.

This treatment is effective in laying the dust for about

6 weeks, though disintegration of the surface may begin in 3 weeks.

Col. Spencer Cosby describes the use of oil in the Washington parks as follows:

" All ruts and holes in the surface of the road are first repaired by cleaning out the cavity, filling it with coarse stone, which is covered with a coating of hot, heavy, asphaltic oil, then sprinkling a light coat of screenings over the oil and finally compacting the mass by ramming. When all holes have been repaired, the surface of the road is thoroughly cleaned with rattan brooms, care being taken to remove all loose materials and caked dirt or dust so that the stone forming the wearing surface of the road shall be exposed and clean. When the road is entirely free from moisture, and during warm, dry weather, if possible, a light asphaltic oil is spread without being heated over its surface by means of sprinkling wagons. One-third to one-half gallon of oil to the square yard usually forms the first application. To allow it to penetrate into the surface, the road is closed to traffic for at least 48 hours after the first application. At the end of this time the surface of the road is covered with a thin coating of clean, coarse, sharp sand or broken-stone screenings, free from dust; it is then rolled and traffic allowed to go over it. A cubic yard of sand or screenings usually covers from 75 to 125 square yards of road surface. In this climate and under the conditions of traffic obtaining on our park roads, the oiling treatment described above keeps the surface in excellent condition for a year. It is never dusty and is muddy only when for a few hours after a heavy thaw the skid chains of automobiles tear up the surface. The subsequent passage of automobiles without chains soon irons out the roadway. At the end of the year the surface of the road is again thoroughly cleaned, from $\frac{1}{4}$ to $\frac{1}{6}$ of a gallon of oil to the square yard under normal conditions is spread over it, and the road closed for 48 hours and covered with sand or screenings as before. This treatment is continued from year to year.

" Instead of handling the oil in barrels, we have found it much cheaper to buy it delivered in tank cars, from which it is unloaded into the sprinkling wagons. A pressure-tank wagon was used to advantage for the first application of oil to the road surface, but ordinary sprinkler wagons with an oil-distributing attachment and a squeegee fixed behind the distributor were found more economical and equally efficient in spreading the oil the second year. To insure coating all parts of the road with an oil layer of uniform thickness, men with stiff brooms followed the sprinkler."

6. *Surfacing macadam pavements with asphaltic oils, asphalts, coal-gas tars and water-gas tars by one application to form a surface* which endures for at least one year.

The application of the bituminous material must be preceded by a thorough cleaning of the macadam road which is to be treated. Pavements whose surfaces are of the larger sizes of broken stone offer a good bond for the bituminous application, which may be made by hand or by gravity or pressure distributors. A type of American pressure distributor is shown in Fig. 37, page 121.

Pressure distributors seem to have an advantage over gravity distributors in that by their use a better bond is secured between the macadam and the surface application. The explanation seems to be found in the washing away of the dust film coating the macadam surface due to the high velocity of the bituminous material as discharged from the nozzles of the distributor.

The amount of tar or asphalt necessary is about $\frac{1}{4}$ to $\frac{1}{2}$ gallon per square yard.

On top of this must be placed a thin layer of sand, gravel or screenings, from 7 to 35 pounds per square yard. In England the top dressing has been omitted, but in this country it is generally specified, since by its use the road may be much sooner opened to traffic. Though rolling is not absolutely necessary, it greatly improves the surface. Fig. 38 shows a type of road roller used in the Boston parks.

For parks with narrow, winding, and steep drives or paths a tandem roller is to be preferred. Fig. 16, page 49, shows a special tandem roller built for this purpose which combines the short wheel base and narrow tread of the tandem roller with the high compression given by the three-wheeled roller.

7. *Impregnating Earth Roads with Crude Oils.* This method originated in California, where a high-grade asphaltic oil is plentiful. The process consists in loosening the soil for a depth of 6 inches. Heated oil is then applied and thoroughly kneaded into the soil by means of a "rolling tamper." The work can be done only in warm dry weather. Porous soils are better adapted to this treatment than heavy ones. Three hundred barrels per mile to be applied once each year are sufficient for a 12-foot roadway.

8. *Impregnating New Macadam Pavements with Bituminous Materials by the Penetration Method.* There are several methods employed, for which see *Highway Engineering*, by Blanchard and Drowne, to which work the author is indebted for much of the information here presented on pavements. One of the best is as follows:

The upper course is built of crusher run from $1\frac{1}{4}$-inch to $\frac{1}{2}$-inch size. On this without further filling of the voids

Courtesy Austin-Western Road Machinery Co.

FIG. 37.—Pressure Distributor.

Courtesy Buffalo Steam Roller Co.

FIG. 38.—Pavement Repairs at Boston.

the bituminous material is applied hot. The rolling ma be done either before or after this application. Som times if the rolling is done afterwards, the rolls are apt pick up the surface. If they are wet or oiled this difficul will be avoided. About $1\frac{1}{2}$ gallon of bituminous ros material is required for each square yard. This shou be applied in dry weather only, preferably when the weath is warm, and it must be applied uniformly. The bitum nous materials used are asphalts, heavy asphaltic oil refined water-gas and coal tars and various combinatio of these materials.

General Remark. The patrol system of maintenanc which in municipal and state work has not been found successful in America as in Europe, owing to government conditions, especially commends itself for parks where well-organized force is always available. Where tl bituminous material exudes to the surface there shou be an application of sand or screenings. Disintegrat spots should be removed and filled with a mixed bituminou aggregate, followed by rolling.

9. *Bituminous Concrete.* The foundation layer shou be 4 to 8 inches in depth of broken stone or telford. Lincoln Park, Chicago, the bituminous concrete was la as follows by Mr. Arthur S. Lewis. The bituminou mixture consisted of $\frac{1}{2}$-inch limestone, torpedo sand ar building sand to which asphaltic cement to the amou of 8 per cent. by weight was added. It was applied in layer $2\frac{1}{2}$ inches deep and rolled with a 10-ton tandem roll by which the bituminous layer was compressed about inch. After the rolling a squeegee coat of pure aspha was applied, using $\frac{1}{2}$ gallon per square yard applied imm diately after the rolling. On the squeegee coat was sprea a thin layer of granite screenings and the rolling was r

repeated. This pavement for boulevard construction is reported to be giving excellent satisfaction.

A large number of mixers are on the market for hot mixing of the materials of both portable and stationary types.

There are several proprietary bituminous concrete pavements before the public which while giving good results will not be discussed in this connection.

The Brick Pavement. This is a type of pavement well adapted for park entrances and plazas where driving is heavy. It is practically dustless and very durable.

The foundation of a modern brick pavement consists of 4 to 6 inches of concrete on which is placed an inch of sand, clean, moderately coarse and free from loam and pebbles. The sand is leveled off with a template, which gives it a true surface. The pavers are required to stand on the brick already laid.

The sand filler has been abandoned and the joints are now filled with cement grout, coal tar, asphalt or a patented filler of bituminous nature.

In order to provide for expansion and contraction a thin strip of wood is placed between the curb and the pavement. This strip is afterward removed and the space filled with a bituminous filler. Similar transverse joints are placed from 25 to 50 feet apart.

The Concrete Pavement. This type of pavement might well be employed for entrances or plazas or wherever the driving is heavy. Its appearance is to be preferred to that of the brick pavement for park purposes. The author believes that the concrete pavement will soon come into favor with park authorities. There are two methods of construction, the Mixing Method and the Grouting Method.

The Mixing Method. A two-course pavement is the more common in municipal practice and will be briefly described here. On the finished and rolled subgrade is deposited a layer of 1 to $2\frac{1}{2}$ to 5 concrete which should be 4 inches thick after tamping. The wearing course may be made of concrete of the proportions one part cement, one part sand and one part $\frac{1}{4}$- to $\frac{1}{2}$-inch granite or trap chips. The wearing course should be 2 inches thick. The leveling off of the wearing course is done with a wooden template resting on scantling set true to line and grade. Transverse joints for expansion are made every 25 feet.

The Grouting Method. The method used by the Hassam Paving Company will be briefly set forth. On the rolled subgrade is placed a layer of broken stone ranging from $1\frac{1}{4}$ to $2\frac{1}{2}$ inches and rolled down to a thickness of 4 inches. This is grouted with a 1 to 3 grout. The surface layer consists of a 2-inch layer of crushed trap poured with a 1 to 2 grout. On the surface layer is thinly spread a thick grout consisting of 1 part cement, 1 part sand and 1 part trap screenings the size of a pea. This is broomed into the surface with stiff brooms. The Long Island Motor Parkway pavement was constructed by the Hassam method.

Concrete pavements are apt to be slippery in winter. To prevent this they have been covered with a bituminous coat in which is incorporated small trap rock chips.

Curb. Curb may be of bluestone, sandstone, granite or concrete. In order to preserve the alignment of a stone curb a concrete foundation is used, the sand foundation having become practically obsolete. The setting of the curb should take place before the pavement is constructed.

Straight curb should be carefully aligned. It is the custom to set a piece of curb every 25 feet by aid of the engineer's measurements and then stretch a mason's line between, thus determining the line and grade of all intermediate pieces.

Brick drives must be lined with curb. This is generally flush with the surface of the pavement. A hollow vitrified-block curb and drain has recently been used which is also flush with the surface of the pavement, which it holds in place.

LIGHTING

The proper methods of park illumination together with practical hints which should prove of value to the inexperienced park engineer are presented on pages 72 and 73. Detailed descriptions of electric, gas, acetylene and gasoline lighting may be found in works on Illuminating Engineering.

INDEX

Printed in Dunstable, United Kingdom

91220030R00087